OUACHITA TECHNICAL COLLEGE
LIBRARY/LRC
3 9005 00015 5634

The Parrot's Theorem

The Parrot's Theorem

A Novel

DENIS GUEDJ

Translated by Frank Wynne

THOMAS DUNNE BOOKS
St. Martin's Press New York

THOMAS DUNNE BOOKS
An imprint of St. Martin's press

www.stmartins.com

Library of Congress Cataloging-in-Publication Data

Guedj, Denis.
[Théorème du perroquet. English]
The parrot's theorem: A novel / Denis Guedj ; translated by Frank Wynne.
p. cm.
ISBN 0-312-28055-6
I. Wynne, Frank. II. Title.
PQ2667. U3555 T4813 2001
843'.914—dc21 2001034891

First published in the United States by St. Martin's Press
First published in Great Britain by Weidenfeld & Nicolson
First published in France by Editions du Seuil, Paris
under the title *Le Théorème du Perroquet*

First U.S. Edition: September 2001

10 9 8 7 6 5 4 3 2 1

Contents

For Bertrand Marchadier

Thanks to Brigitte, Jacques Binsztok, Jean Brette, Christian Houzel, Jean-Marc Lévy-Leblond, Isabelle Stengers.

CHAPTER 1

A Bird in the Hand

A curious parrot; a Euclidean family & two friends called Being and Nothingness

Max Liard set out over the hill to the flea market at Clignancourt, as he did every Saturday. He rummaged for a while on the stall where his sister had traded in the Nikes that Perrette, their mother, had given her, and then wandered into a huge warehouse that sold army surplus. As he was ferreting about in a pile of clothes, he saw two guys at the back of the shop who seemed to be arguing. It looked as if they were fighting. None of his business, really. But then he saw that the two men were trying to catch a parrot.

The parrot put up a good fight. The shorter of the two guys caught it by one wing, but it wheeled round and bit hard into his finger, drawing blood. Though he heard nothing – Max was deaf – he *saw* the guy scream. The bigger man lashed out, hitting the parrot squarely between the eyes. Max moved a little closer. Though he knew it was impossible, he thought he heard the parrot shouting 'Help!' One of the guys took out a muzzle and tried to put it over the parrot's head. Max decided that this was his business after all, and ran towards them.

The rue Ravignan is short and steep, running from the fountain at the Place Émile-Goudeau, where the Bateau-Lavoir – Montmartre's famous studio of painters – still stands, to the junction of the rue des Abbesses and the rue d'Orchampt. Right in the middle is a bookshop called A Thousand and One Pages, owned by an old man named Mr Ruche. It was more spacious than the other shops on the street, and that's how Mr Ruche liked it. He didn't like cramped spaces. Nothing made him angrier than people who jammed books tightly together on shelves. Books shouldn't be packed together like people crammed into the métro in the rush hour, but neither did he like to see them lolling about, with gaps you could drive a taxi through. One of Mr Ruche's guiding principles was that books needed room to breathe, a lesson he had taught Perrette, the woman who worked with him. Since the accident that had left Mr Ruche in a wheelchair, Perrette had been in sole charge of the bookshop. She was there from dawn to dusk, dealing with customers, orders, deliveries, returns and accounts. She did everything and she did an excellent job.

This morning the stench of motor oil on the rue Ravignan was so strong that Perrette had to hold her breath as she walked through the old garage that was Mr Ruche's bedroom. She pushed aside the curtains and handed Mr Ruche a letter. From the large Brazilian stamp in the corner and postmark, Perrette knew it had been posted two weeks earlier, in Manaus. Mr Ruche didn't know anyone in Brazil, let alone in Manaus.

Monsieur Pierre Ruche
1001 Pages
Rue Ravignan
Paris XVIII, FRANCE

It was certainly addressed to him, but there was no street number and the name of the shop was strange: '1001' instead of 'A Thousand and One'.

*Dear πR**

From the way I've spelled your name, you should be able to guess who I am. Right first time. It's me, Elgar – get your breath back. I know – we haven't seen each other for half a century. I've counted.

The last time I saw you was the day we escaped, do you remember? It was 1941. You told me you were leaving to fight a war you had nothing to do with starting. I wanted to get out too; to get away from Europe and from a war that I thought had gone on too long. And I did. After we left each other, I took a boat to the Amazon rainforest. I've been living here ever since. I live near Manaus. I'm sure you've heard of it – it used to be the rubber capital of the world.

*The name Pierre – Peter – is pronounced 'pee-air', rather like πR, the pronunciation of πR in French (as in the equation πr^2).

Why write now, after all this time? Because I've sent you some books. Why you? Because you were my best friend in the whole world, and anyway you're the only bookseller I know. I'm sending you my library: all of my books – almost a ton of books about maths.

All the classics are there. I suppose you think it's strange that I refer to maths as if it were literature, but I guarantee that there are better stories in my books than in the best novels. Stories of mathematicians like 'Umar al-Khayyām or Nasīr al-Dīn al-Tūsī, the great Italian Niccolò Fontana, known as Tartaglia, and the French mathematician Pierre Fermat. And those are just a few off the top of my head. You might not agree with me. I know lots of people think that knowledge is just a strange collection of facts tied up with an education. But, if you ever decide to read one of the books, do me one favour: before you start, ask yourself, 'What story is it trying to tell?' If you do, I'm sure that you'll see the world of maths you thought was difficult and grey in a different light. It might even make a confirmed reader of novels like you happy.

In the crates I've sent, I think you'll find the world's best collection of books about mathematics. Everything that should be there is there. It's certainly the most complete private collection on the subject ever made.

How did I come by it? A bookseller like you will wonder how much it cost me. Not just in money, but in time and energy. I can tell you, it cost a fortune. There are first editions in there, some more than five hundred years old, which took me years to track down. How could I afford them? I think I should draw a veil over that. It hasn't always been by hard work and honest means, but I can promise that not one has blood on it – though a few might be stained with a drop of alcohol, or a difficult compromise.

The books that I have collected were for me alone. Every night, I'd choose one to spend the night with; and they've been wonderful nights, torrid, humid equatorial nights. Nights every bit as important to me as when we were sowing our wild oats long ago in the hotels around the old Sorbonne. But I'm getting sidetracked.

One last thing. Unless you've changed beyond all recognition, I'm certain of two things. Knowing how little money motivates you, I'm sure you will never sell these books. And knowing how little maths interests you, I imagine you will never read one of these books. Consequently, I feel confident that they will come to no harm in your care.

Love,

Your friend, Elgar

This was a deliberate taunt, which proved to Mr Ruche that Elgar Grosrouvre hadn't changed one bit. Mr Ruche vowed that this time he would get one over on his old friend: he *would* read the books, and he *would* sell them. (This, of course, is precisely what Elgar intended. For his plan to work, Mr Ruche had to read the books first, because Elgar knew that once he had, he wouldn't be able to part with them.)

The Amazon! Why on earth had Elgar gone to the Amazon, and what was he doing in Manaus? Mr Ruche was so engrossed in thought that he didn't even notice the postscript on the other side:

PS The packages I had made up have all come apart. I've had to repack the books higgledy-piggledy into packing crates. You'll have to sort them out however you see fit, πR. It's not my problem any more!
PPS I may come and pay you a visit. Since we're not getting any younger, it would have to be quite soon. Will you recognize me? My hair is grey now and my face is fat – my feet are red as lobsters from the heat. I think I've turned into an old wizard.

The bookshop and the garage room took up the ground floor of a two-storey building that ran up the steep escarpment of the street. A narrow hallway between them led to a small courtyard. On the first floor was a flat where Perrette lived with her children. A galley kitchen ran along one wall of the living room; another wall was dominated by a massive chimney breast. The windows on the first floor opened onto a long balcony from which a narrow staircase ran down to a courtyard where an ancient bay tree stood. On the west wall was a fountain with a broken tap that dripped continually into an oriental basin. There were two artist's studios, now empty. In all, the place had a Moroccan feel to it.

Perrette's room had once belonged to Mr Ruche. Max, her youngest son, had a room sandwiched between the narrow toilet and the large bathroom. The attic had been converted and was divided into identical rooms for the twins, Jonathan and Lea. The rooms were bright, lit by large skylights in the sloping tiled roof, and at night gave a beautiful view of the city. Up in their eyrie Jonathan and Lea were like astronauts, looking out onto the sky, the clouds, the stars. The two cheap plastic skylights connected them to the infinity of space.

In the courtyard was the lift, which Mr Ruche had had installed after the accident. He got the idea from the barrel lifts that were to be found in every café in Paris. Hidden under a trapdoor behind the bar, they were used for lowering barrels of beer into the cellar. On the rue de Ravignan, they had no need to lift barrels. This lift was to take Mr Ruche from the courtyard up to the first-floor balcony. All he had to do was manoeuvre his wheelchair onto the platform, apply the brakes and push the button. It made for a strange sight – an old man in a wheelchair, rising into the air, on a lift shaded by a multicoloured umbrella!

Mr Ruche slept on the ground floor in the garage he had had converted into a bedroom. He couldn't use his car any more – and it would have reminded him of when he used to take long drives along the country roads. Being at ground level made it much easier for him to take his daily constitutional – something he couldn't live without. In the summer he could sometimes smell motor-oil from the floor, and it brought back memories.

He had furnished the room extravagantly with a four-poster bed and yards and yards of purple velvet draped on every wall. Mr Ruche said it was 'a king's bedroom for a pauper'. In the corner of the room was a dresser filled with shoes – little worn since the accident – and on it was a sign:

> *It is impossible to understand the science of shoes*
> *until one understands what science is* (Plato, *Theaetetus*).

For some time now, Mr Ruche had lived quietly, expecting little out of life, happy to drift through a quiet retirement. Now, suddenly, a letter from halfway round the world looked set to change all that.

He remembered the first time he had met Elgar Grosrouvre. They were in their first year at the Sorbonne: Mr Ruche was reading philosophy; Grosrouvre, maths. While there, both began to write. Ruche published a much-admired essay on *being*; Grosrouvre wrote a highly acclaimed thesis on the number zero. They were inseparable – other students jokingly called them 'Being and Nothingness'. When Sartre published his essay some years later, Mr Ruche suspected he had stolen the name, but he couldn't prove it.

Mr Ruche lifted himself from the bed to his wheelchair and set off on his daily constitutional. What could Grosrouvre possibly want? Perhaps his old friend was trying to rattle him, to make sure he didn't get too comfortable in his old age. Was this a present or a time bomb?

When he got back home, he phoned the carpenter on the rue des Trois-Frères and began to plan the conversion of one of the empty studios into a library to house Grosrouvre's books, if they ever arrived. He wasn't sure that they would, but when Grosrouvre said he would do something, he generally did. A ton of books could arrive any day. Even if they didn't arrive, he could use the converted studio as a stockroom.

His nose scratched, his ear battered, his left hand bruised and his trousers torn, Max pushed open the door to the living room. At eleven years old, Max knew a bargain when he saw one and always came back from the market with something strange and valuable. This time, it had feathers and it stank. The bedraggled parrot was perched on Max's good hand. He set the bird down on the back of a chair, near the coffee table where his brother and sister, Jonathan and Lea, were finishing their breakfast. They stared at the bird.

The parrot stood just over a foot tall, rocking on its feet. Its green feathers were coated in a layer of dust through which it was just possible to make out that the tips of its wings were bright red. Its most striking feature was the blue feathers on its forehead, though there was a nasty cut. The bird could barely keep open its great dark eyes.

The first thing to do was wash the bird. The parrot seemed indifferent and let Max clean its feathers and feet, but when he tried to clean its beak the parrot's eyes lit up angrily for a brief moment. Though the bird looked as if it might topple over from exhaustion, it summoned up the energy to fly a short distance, landing on the cornice above the chimney breast. There it fell asleep, its head back and buried in its feathers.

No one noticed Perrette come in. She had just come back from the hairdresser, where she'd had her hair permed. She was beautiful but clearly didn't think her looks important. Her hair was short and curly, and she wore little make-up. She was clearly not in a good mood. 'This place smells of cat's piss!' she said.

'A parrot can't possibly smell of cat's piss,' corrected Jonathan, 'though it does stink!'

'I suppose,' said Lea, 'it could smell of parrot's piss.'

'Parrot? What parrot?'

Perrette looked around, and the twins pointed to the parrot perched on the cornice.

'Get that thing out of here!'

'But, Mum,' said Max, 'he's asleep!'

'Maybe we could wait till he wakes up', said Lea, though she wasn't keen on keeping the parrot.

'This place is bad enough already what with a pair of delinquent twins, a deaf son and a cripple. The last thing we need is a parrot.'

She was so angry that she hadn't heard the squeaking of the wheelchair behind her. She blushed. 'I'm sorry, Mr Ruche' she said, eventually.

'Nothing to be sorry about. It sounded quite an accurate description of the household.' Mr Ruche had noticed that Perrette had been on edge for several days now.

'I like your new hairstyle', he said, making little circles with his finger.

She looked at him, mystified. 'What?' She ran her fingers through her hair. 'Oh, of course. I think they overdid the curls, though.'

'Mother, I think I should explain,' said Jonathan, and he told his mother the story of how the parrot came to be there. When Jonathan told her about the fight, she wheeled round. She hadn't noticed the scratches on Max's face, or the bruising. She went over to look at him. It didn't look too serious.

'What do you think about all this, Mr Ruche?'

'I don't think he'll have any scars.'

'Not Max, about the parrot?'

'I think the parrot may well have a scar.'

'No, I meant should we keep him, or...'

'Oh, I think if we throw him out on the street after all he's been through it would be a clear case of parrot-abuse.'

They all laughed. Except Max. He looked at his mother for a minute and then said, calmly, 'You wouldn't really refuse a parrot in distress, would you, Mum?'

'Does it talk?'

'Not a word...at least not since he's been here', Max assured her.

'Well, I suppose we can grant him a temporary visa.'

Stretched out on their beds, each under their own skylight, Jonathan and Lea talked to each other through the half-open door between their rooms.

'Why would two men – "big guys" according to Max – want to put a muzzle on a parrot?' asked Jonathan.

'To stop it from talking, I expect', Lea answered.

'You think it was talking rather than biting?'

Between them, the twins were thirty-three years old and ten foot six. Jonathan was the elder, Lea the younger (by two and a half minutes). Everyone called them Jon-and-Lea. She always seemed to be trying to catch up on those two and a half minutes. She wanted to be first in everything, and most of the time she was. Jonathan seemed happy just to have been the first-born. They were as alike and as different as it was possible to be. It was as if they were the same person in different wrappers. But their eyes were the same – pale blue, like faded denim.

Lea was already a young woman. She wore no make-up, but dyed her cropped hair a different colour every week. Perrette told her that the bleach would ruin her hair, but she took no notice. She always wore a T-shirt and jeans with trainers or a pair of Doc Martens. Lea was lithe like a jungle creeper and thin as a rake – Euclid would have said she had 'length but no breadth'.

Jonathan had long, curly hair. He wore baggy clothes and a gold stud in his right ear. He radiated health and never seemed to feel the cold. He'd had acne as a boy, but now he had just a single spot on the tip of his chin. Jonathan scratched it sometimes when he was upset. He had beautiful hands. He wasn't fat, but stocky – Euclid would have said he had 'both length *and* breadth'.

If the twins were length and breadth, Max was the deep one in the Liard family. He was chubby with a wide forehead and a mop of copper-red hair. He had small, dark eyes like lumps of coal. If he frowned his eyes all but disappeared, but mostly they glittered. He was very muscular for an eleven-year-old. He had a seriousness which was surprising in a boy his age and often made people uncomfortable. He was assured and self-confident. What would

Euclid have said about Max? That he was a *solid*. He had 'length, breadth and depth', so he must be a solid, though you wouldn't think it from the way he flew about the place.

How had Max been able to lip-read a parrot calling for help? He hadn't, but he'd understood all the same. Max thought that sounds were like icebergs. What you heard was like the ice above the surface – most of the sound was beneath. He had developed a sort of sixth sense: it was as if he listened with his whole body and heard things that went unsaid. Mr Ruche had spotted his talent, and called him Max and Zephyr because he could sense things in the air.

CHAPTER 2

Seen and Not Seen

A floating library & a family mystery

The rain lashed hard against the windows as the ship lurched and heaved. The captain, Bastos, had been at the wheel for hours now and he was exhausted. He was sweating, in spite of the cold, and the radar seemed to be on the blink. The door to the bridge banged open and his first mate was thrown inside, grabbing for the rail to steady himself. He looked scared.

'I checked the hold, Captain. The cargo seems to be safe, though I don't think we can take much more. Two or three more waves like that last one and the ropes will break. We're carrying too much weight, Captain.' He cleared his throat. 'If this keeps up, we'll have to jettison the cargo.'

Bastos turned and yelled, 'We will not! In all my years as Captain I've never lost a single crate – not one. My father worked this route and his father before him. I won't have any talk of jettisoning the cargo. Now, make yourself useful and see how things are in the engine room.'

The first mate hesitated, about to speak.

'That's an order!'

They had left Belem, on the coast of Brazil, three days before. Bastos knew this route like the back of his hand. In the thirty years he had been making this crossing he had never seen a storm like this. He knew the ocean well, but the savagery of the elements surprised even him. He had the best crew in the South Atlantic. He had hand-picked every one of them, each tough and experienced. His first mate was a fine man, and they had worked together for years. Bastos tried to remember what the cargo was. Some furniture and a dozen containers, and a crate of books from Manaus.

There was a sudden terrifying silence as the engines cut out. After a long moment – it seemed like an eternity – they roared into action again, but they sounded weaker now. The boat had slowed to a crawl. One of the engines had given out.

There was nothing to do but jettison the cargo. With one engine down they were pitched like a toy on colossal waves that could dash the ship to pieces. If the storm did not break, the ship would go under. Bastos had his men to think about and he had to think fast. He had no intention of going down with his ship; he would tell the men to jettison the cargo and pray to God that it would be enough to save them all.

There was a terrible crash. The boat soared from the water as if it had wings. It rose to the crest of the wave from where, through the fog, Bastos thought he could make out a ship bearing down on them.

Back on the rue Ravignan, the parrot hadn't moved from its perch. It looked like an inert ball of feathers. Max set up a stepladder beside the chimney breast, climbed up and sat on the top step. He reached out towards the parrot but then stopped. He didn't think it would be right to stroke the bird while it was still asleep.

'Come on...come on boy, time to wake up. I know you can talk, I saw you in the warehouse but you haven't said a word since you got here. I'm deaf and you're dumb – we should get on like a house on fire. You'll be safe here, you just concentrate on getting better. I saw that bastard hit you. If I get my hands on them...I expect they're looking for you. Let them. They'll never find you in a city the size of Paris.'

Max turned to make sure that no one had come in. 'If I'm not looking, I can't lip read, so I don't understand anything. You can't imagine what it's like being deaf. Nobody can. The only voice I hear is my own going on and on in my head all the time. Sometimes I feel like I need a holiday from my own voice. The twins never feel like that. Did you see them? There's two of them, but it's like they're one person – Jon-and-Lea – like it's just one word. I'm Max – Max the Zephyr, Mr Ruche calls me. Oh, and I talk too much. I'm lucky I wasn't born deaf otherwise I wouldn't be able to talk at all. If you can't hear, you can't talk. You can make sounds only if you can hear them. It's not just words I miss, but the noise of the fountain in the courtyard or Mr Ruche's wheelchair. I can do them all, listen!'

In a low voice he made a sound like the fountain running in the courtyard, then the sound of Mr Ruche's wheels squeaking. 'See? We're all parrots, really, we just repeat the sounds we hear.' He laughed, and the stepladder wobbled. Max grabbed the cornice to steady himself.

'The only things we don't copy are screaming and crying. You don't need to hear them to be able to do them. And laughing, maybe, but I'm not so sure.'

The Liard family sat down to lunch at the long dining table. Lea was twirling a mountain of spaghetti with two forks to mix in the sauce as her mother watched. Suddenly a hoarse voice croaked, 'I won't say anything without a brief.' It was the parrot.

Though Max had his back turned, he knew from the faces of the others that something had happened. He turned round. The parrot was preening itself: its feathers gleamed, the wing-tips glowed bright red and in the centre of its blue forehead was a thin red line where the wound was beginning to heal.

Perrette was the first to react. 'You said he didn't talk!'

'Well, he's talking now,' said Jonathan, 'but only to say that he won't say anything.'

'No. He said he won't say anything without his brief present – he means his lawyer', said Mr Ruche.

'Pretty strange thing to say, if you ask me', said Lea.

'He said it because he heard someone else say it!' said Jonathan. 'He's just repeating.'

'Maybe his owner was a lawyer?' wondered Lea.

'No,' said Max, 'he must have been a criminal. That's what criminals say.'

'Maybe that's what he was saying to the guys in the warehouse, Max, when they were trying to do him over.'

'They weren't trying to do him over, they were trying to muzzle him', Max corrected Jonathan.

Perrette burst out laughing. 'You've all been watching too many cop shows', she said. 'He didn't say a *brief*, he said a *brie*. The poor bird is starving and he wants some cheese.'

Max popped out to the corner shop, but Mr Habibi didn't have any cheese, so he had to go to an all-night supermarket near the rue de la Goutte d'Or. He came back with a whole brie weighing more than two pounds, and the parrot gobbled it up.

Though the blow to his head had healed, the bird didn't seem to remember anything. It put him in the unique position of being the only parrot to repeat things it could not remember hearing. They called him Sid Vicious – Sidney for short – because the multi-coloured feathers of his plume made him look like a punk with a mohican.

They set up a perch in the dining room with a feeding bowl and a small bird bath, making sure he was out of the draught. Max put a tray under the feeding bowl to catch any crumbs. The parrot quickly learnt from Max that his name was Sidney.

Her son's question, 'You wouldn't really refuse a parrot in distress, would you, Mum?' had surprised Perrette. She decided it was time to tell the children everything. It was time that they knew how the five of them came to be living here together on the rue Ravignan.

It had all started with a fall seventeen years ago. Perrette was almost twenty then; she was studying law and was engaged to be married to a young lawyer. She had met him on a skiing holiday in the Pyrénées. He invited her to go on holiday with him on the Côte d'Azur later that year. While they were there, he proposed. They were to be married in Paris in the summer.

A month before the wedding she was on her way for the final fitting of her wedding dress, and was so caught up in her own thoughts and all the little things still left to do that she never even saw the open manhole. Perrette screamed as she felt the ground open up and swallow her. Nobody saw her fall. It was two hours before she was rescued. By the time she got to the wedding outfitters, the shutters were down and the door was closed. She limped back to her flat, unplugged the phone and ran a bath. The following day she called off the wedding. She never gave the slightest explanation, and her parents never forgave her the humiliation. She cut herself off and never saw her parents or her fiancé again. Nine months later Jonathan and Lea, her identical-but-not-identical twins, were born.

She found a job at Mr Ruche's bookshop, A Thousand and One Pages, and after the twins were born, he suggested that she come and live in the house on the rue Ravignan. She accepted immediately. She was happy in her job, but she decided she wanted another child. Against the advice of social workers she adopted Max, who was six months old when he came to live with her and the twins in the flat above the bookshop.

That evening, when she had told them her story, no one dared speak. Everyone she cared about was there: Max, Jonathan, Lea, Mr Ruche. This was her world. They had listened carefully to every word she said, as secrets she had kept for seventeen years came tumbling out in a few breathless sentences. Now she had said it all, in a flat voice, not daring to look them in the eye. In one moment their whole world was turned upside down; they knew something more about themselves. Mr Ruche had always known.

It was a relief for Perrette. She had never talked about her fall before, never breathed a word about adopting Max. Mr Ruche, who was the only one who knew something, had never questioned her.

She looked up and said to Max, 'You're not mine, but I chose to keep you'.

To the twins, 'You two *are* mine and I chose to keep you.'

Then to her three children: 'You are all mine and I am yours.'

She lit a cigarette. Mr Ruche held out his hand.

'Could you spare me a cigarette, Perrette?'

He hadn't smoked a cigarette in years. Perrette held out the pack and he took one; as she held the match to the cigarette she said softly to him, 'And you, Mr Ruche, have given us a home.'

After a while she stubbed out her cigarette and got up, a little unsteadily, trying to look calm. Suddenly a smile lit up her face. 'Goodnight, everyone', she murmured, and she seemed to float out of the room like a feather.

Perrette slipped into bed, and for a second she remembered the fishmongers on the corner of the rue Lepic. Every time she walked past it, she silently thanked the owner for refusing to hire her when she most needed work. How different their lives would have been if she had been selling mackerel and mussels instead of books. Then she fell asleep.

In the living room, Max, wearing his pyjamas, was leaning against Sidney's perch. The parrot's eyes glittered in the half-light. He was listening carefully as Max spoke. 'I don't know where you're from,' Max said, 'and it doesn't matter because now I don't know where I'm from either. You heard what Mum said: "I chose to keep you."' He stroked the bird and it arched its neck towards him. 'And I chose you', he said. 'And I'll make sure it's not a temporary visa.' He smiled. 'I knew that from the moment we left the market.'

Under the attic skylights, Jonathan stared out at the starless sky and the clouds washed orange by the city lights, and decided to ask the question that had been burning inside him. 'What do you think she meant when she said "Nine months later...?"'

'...the twins were born', Lea interrupted. 'Surely I don't have to draw you a picture? She meant we were born in the gutter.'

'No!' snapped Jonathan, 'that we were conceived in a gutter.'

Lea could tell he was angry. 'I suppose you would rather we were conceived on a big, soft bed with silk sheets and flowery pillowcases. I suppose you'd rather our dad was a nice clean young lawyer? You're so predictable!' she sounded disgusted.

'What I'd prefer' – Jon was furious now – 'is if she said "I'd rather not tell you anything about your birth" instead of coming out with something as ridiculous as that. I'd rather', he spat, 'she told us the truth!'

'She did tell us the truth!'

Downstairs in his four-poster bed, Mr Ruche was muttering to himself. 'It's all happening too quickly. Grosrouvre's books, Perrette explaining everything and this parrot...what do they call him? Sid Vicious. I wonder why Perrette never said anything before? Why wait all this time? I should talk to them – the twins especially. But what do I say? I've no idea how to talk to children. Worse, they're not even children – they're teenagers. I have to think of something.'

He fell asleep before that something came to him.

CHAPTER 3

The Glass Pyramid

Thales and the beginnings of mathematics; Mr Ruche lives dangerously & a visit to the pyramids

'In the reign of King Gugu, on the shores of the Aegean sea, Thales, son of Examius, was walking in the country near Miletus, a small town in Ionia.'

Jonathan woke up. How dare someone wake him so early on a Sunday? He opened one eye, like a bulldog, and scratched the spot on his chin. The door between his room and Lea's was open as usual. The high nasal voice kept talking: 'Thales walked across the fields with a servant-girl by his side.'

It wasn't Lea – it had to be the radio.

'And as he walked, he studied the skies.'

It wasn't the radio. Jonathan leapt out of bed and ran to the door.

'I must be hearing things!'

It was the parrot, perched on the door frame. Lea came out of her room and was just as surprised to find the bird there, about to launch into a new speech. They stormed down the stairs. It was eleven o'clock. Max was putting away the breakfast things and Mr Ruche was pretending to read his newspaper.

'I suppose you think it's funny,' Lea shouted, 'sending a parrot to wake us up first thing on a Sunday morning? He was outside our door repeating some rubbish you've taught him.'

With a flap of its wings, the parrot flew past Lea and squawked, 'I don't repeat, I don't report, I don't relate, I don't parrot. *I'm telling you!*' The feathers round the scar on its forehead stood out like spikes. This was an angry parrot.

Jonathan played with his earring. 'Isn't it a bit early to be going on about Thales?' he asked. 'We haven't even had breakfast yet.'

'As Sidney was about to tell you – in his words not mine – Thales was studying the sky to see if he could discover something of the secrets of the movements of the planets. The servant-girl walking beside him noticed a hole in the ground just in front of her and walked around it. Thales, who was busy studying the sky, fell in. She said, "If you can't even see what's in front of you, what makes you think you can see what goes on in the heavens?", offering him her hand to help get him out. Everything', concluded Mr Ruche, 'starts with a fall.'

The door opened and Perrette came in carrying the shopping. She had only heard the last sentence. Jon-and-Lea looked at her and went back up to their rooms. They got the message. As she left, Lea couldn't help saying sarcastically, 'and they lived happy ever after and had lots of kids.'

'Wrong, Lea,' said Mr Ruche gleefully, 'Thales had no children, though he did adopt his sister Kybisthos' son.'

Like most pupils, Jonathan had heard of Thales at some point, but his teacher had never talked about the man, only about his work. In fact, in maths that was all anyone talked about. Now and again a name would be mentioned: Pythagoras, Pascal, Descartes, but it might as well have been the name of a cheese or a railway station. No one ever talked about the people, or about what they'd done. It was as if the formulas, proofs and theorems appeared on the blackboard out of thin air, as though they had always been there, like mountains or rivers. Even mountains and rivers hadn't always been there. Sometimes it seemed that theorems had been around for longer than the earth itself. What bothered Jonathan was that maths didn't seem to be *about* anything. It didn't explain the world, didn't add anything to what he learned in history or geography. Jonathan didn't really know what it was about and he didn't really care.

'You were great!' Max told the parrot after the twins had gone upstairs again. 'That told them!' He pursed his lips: '"I don't repeat, *I'm telling you.*" Good for you. That shut them up. You've got a great memory.'

This was exactly what Jonathan was thinking, lying upstairs on his bed.

'For a parrot that can't talk, he certainly learns fast', he said. 'Have you ever heard one talk for that long before?'

Lea didn't answer.

'Do you remember when Mum used to take us to the pet shops down by the river? We used to stand for hours in front of the birds in their cages, and none of them ever said a word.'

'Maybe they just weren't talking parrots', said Lea, but she was thinking about other things.

'He isn't a talking parrot, he's motormouth!'

Lea left him upstairs and went back down to the living room. She stormed up to Mr Ruche, who was waiting for her next question.

'What?' she said angrily, 'What started with Thales falling into the hole?'

She sat down to eat her breakfast. From the kitchen, Perrette listened. Mr Ruche took his time, then he began.

'Thales was the first real "thinker" in history. I don't mean that nobody thought before him; people had been thinking for thousands of years. There were wise men before him: scribes and priests, accountants and storytellers. People recited prayers, counted money, told stories, but Thales did something different. He asked questions to which he didn't know the answers. "What is thinking?" or "What is the connection between what I think and what really exists?" or "What is nature made up of?" It may seem strange now, but until then no one had ever asked those questions.'

Mr Ruche was in his element, talking about philosophy. Jonathan came downstairs wearing a purple dressing gown and sandals. He shook some cornflakes into a bowl and poured on some milk.

'But that's philosophy, isn't it?' Lea asked.

'Yeah', said Jonathan. 'I thought Thales was a mathematician.'

Mr Ruche smiled. He'd got them – hook, line and sinker. He went on, 'In the sixth century BC – which is when Thales was around – it was all the same thing. In fact there were no words to distinguish philosophy from mathematics. Those definitions came much later, and it was later still that they were thought of as being separate disciplines. Nowadays, nobody seems to remember that.'

He didn't want to stop now that they were on the subject of Thales. He knew a lot about Thales, in fact he considered him to be one of the great thinkers, but he would need to do some research into the maths. He would start in the National Library – where he and Grosrouvre had spent weeks at a time when they were students.

There was a problem. To get into the National Library, he would need a card and that meant a long and serious interview with one of the librarians. When he arrived, the librarian wanted to know everything about him: whether he was a professor or a student; the subject of his research; his tutor's name. 'I'm sorry,' she said apologetically, 'but we have to ask everyone these questions.'

Mr Ruche could hardly say to her, 'Look, I live with a woman called Perrette who has a deaf son called Max. He found a parrot that made a speech about Thales and now the twins want to know more...etc., etc.' She would never understand. He gave the librarian a big smile. 'My name is Pierre Ruche', he said. 'I have a bookshop in Montmartre. I'm eighty-four years old and my tutor died in 1944. I never finished my thesis. Since then, I've been working on it alone. My research is, well, personal – I won't be publishing a paper. I'd like to consult some books on Thales and the origins of mathematics in Greece.'

She put up her hand for him to stop. 'That's fine', she said. 'Would you like a card which allows you ten admissions, or entry for a full year?'

'I suppose at my age I should go for the ten admissions,' said Mr Ruche, 'but let's live dangerously – give me a card for a full year!'

He paid and waited while his photo was taken. It was printed out directly onto the card and sealed in plastic. Without even looking at it, he picked it up proudly and thrust it into his jacket pocket.

When he got to the reading room, he handed in his card and was given a small token with a seat number on it. The room hadn't changed at all. In the old days, Mr Ruche used to be light on his feet in the aisles; now, with a wheelchair, things were a little different. One wheel caught the side of a chair as he passed and it toppled over, then he accidentally ran over a folder someone had left lying on the floor, and almost knocked over a rack full of magazines. When he got to his seat, though, it was just like old times. He settled down at his desk and turned on the lamp – it was strange, even in the middle of the day students always turned on the desk lamps. He asked for the catalogue of books in print and was furious to discover that the catalogues and microfiches were downstairs –

and there was no lift. He was about to complain to the head librarian, when he remembered that there was a copy of the *General Catalogue of Printed Books* in the reading room. He found it and noted the titles he wanted and filled out an order form and went to lunch.

He had a sandwich and a glass of wine in a nearby café with some of the regulars. By half past one the café was almost empty, and he sat enjoying the silence for a moment before heading back. He felt like a student again; a little old, but a student nonetheless. He took out the library card and looked at the photo. It was tiny, but the quality was perfect. He hadn't looked at himself for a long time. His eyes were bright and clear, his hair was combed back off his face, and his jaw square, and there was barely a wrinkle to be seen. He smiled. All the wrinkles were on the inside. He put the card back in his wallet.

He stopped at the newsagent across the road and looked at the notebooks. He was very particular about stationery, but eventually decided on a large black notebook with wide margins. Then he took a taxi back to the rue Ravignan.

After a couple of mornings at the library, his notebook was filling up nicely. He sat in a booth in the great reading room and reread his notes.

> 7th century BC on the shores of Anatolia. King Gugu's son has his court at Sardis, capital of the Lydian Empire, but in Ionia, Miletus is one of the first city states. A free city. Thales born about 620 BC. Most famous for his injunction 'Know thyself'. Considered to be one of the seven wise men of ancient Greece and the first to propose general theories on mathematical principles.
>
> Thales is less interested in numbers than in geometric forms: circles, lines, triangles. First to consider the 'angle' to be a thing in itself, making it the fourth element of geometry (the others being length, breadth and volume). He is the first to observe that the opposing angles made by two intersecting lines are equal.

Mr Ruche drew a diagram:

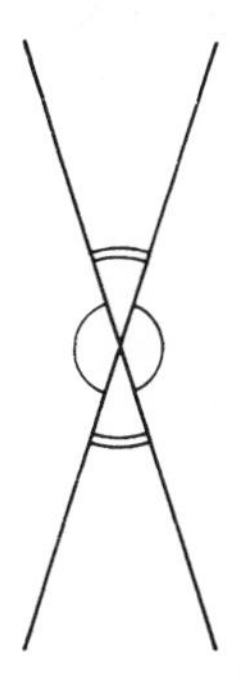

The diagram seemed stark and lonely, somehow. It reminded him how sad he had found geometry when he was young. He carried on reading:

> Link between circles and triangles. Thales showed every triangle corresponds to a circle, which touches all three points, called the *circumscribing circle*, and even made a general rule.

In the margin, he noted, 'This means that one and only one circle will pass through any three points.' He read what he had written. No. That was wrong. He added, 'three points *which are not aligned*', because if the three points were aligned, a circle couldn't pass through them, only a line could. 'Which means', he thought, 'that three non-aligned points define a triangle – that much seems obvious – but also, define a circle – which is less obvious.' As he drew the new diagram, he wondered why Thales was so determined to link the different geometric figures. When he had finished, he took a ruler from his briefcase and drew a frame around it. It made the drawing look less like a diagram and more like a picture:

The girl at the desk opposite looked at him, a little surprised to see this old man drawing shapes in a big notebook. Mr Ruche brushed off the bits of eraser where he had rubbed out a line, then went back to his writing.

> Thales proved that an *isosceles* triangle has two equal angles, and using the link he'd made between length and angles, that meant if two *angles* were equal, two *sides* must be equal.

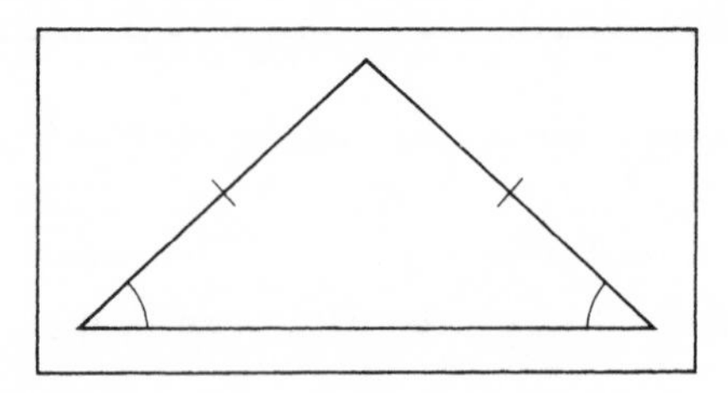

Mr Ruche smiled as he read the next bit.

> Native Americans refer to a bison as a *two-horn*; for a bicycle we say a *two-wheeler* and for a shape with three angles, we say a *tri-angle*, though you could just as easily say a *tri-sider*. In fact that's what a *trilateral* meant in the old days, in the same way they said *quadrilateral*.

Now that he'd started on what *words* meant, he went on:

> Isosceles. *Iso*, in Greek, means 'same', *skelos* means 'legs'. The two legs of an isosceles triangle are the same. Other triangles, which don't have two equal sides, are called *scalene*, which means they 'limp'!

Mr Ruche laughed and imagined a geometry problem beginning, 'If you have a limping triangle...' He thought about Perrette and her isosceles family: two equal children – the twins – plus one.

Having dealt with Thales' discoveries about the links between circles and triangles, angles and sides, Mr Ruche set to work on lines and circles, diving into a book on the origins of Greek mathematics.

As he was jotting down the information, he thought of a line from Grosrouvre's letter: 'I guarantee that there are stories in those books better than the best novels.' Maths as good as Shakespeare or Dickens or Tolstoy. As usual, Grosrouvre was laying it on a little thickly, but Mr Ruche had to admit that it was a new way of looking at maths. Maybe he should take Elgar at his word. What story were these pages trying to tell?

This story takes place on a plane and the characters are a line and a circle, thought Mr Ruche. What could happen to a line or a circle? Well, either the line goes through the circle or it doesn't – though it could also just touch it. If it goes through the circle, then it has to cut it into two. How would you position the line so as to cut the circle into two equal parts? Thales had an answer for this: for a line to bisect a circle, it must pass through the centre of the circle. Such a line is a diameter. The diameter is the longest line that can be contained within a circle. It cuts the circle at its widest point, which is why the diameter can be said to 'describe' the circle.

He picked up his compass and pencil:

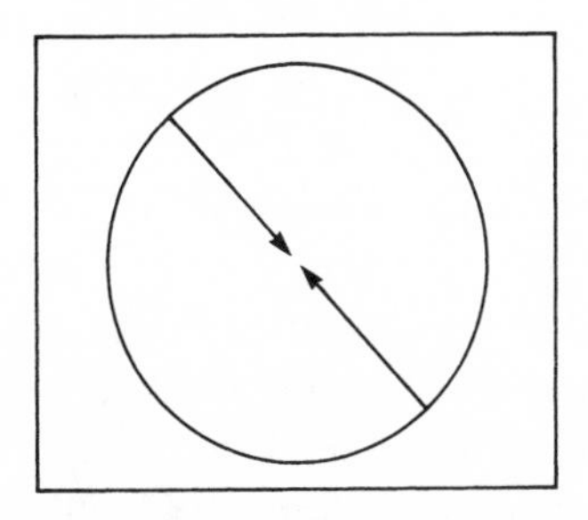

Thales' solution does not apply to one *particular* circle, but to any circle. He was not trying to work out a numerical answer to a particular problem, as the Egyptians or the Babylonians had done before him. Thales wanted to make statements which were true about a whole class of objects. This was a whole new idea. To be able to do this, Thales would have to imagine an ideal object – *the* circle, which represents 'every circle in the world'! This is why he is sometimes called 'the first mathematician in history'. Thales' view is completely original and new. It's impossible to imagine just how daring and new it was to make statements like, 'A line passing through the centre of a circle divides that circle into two equal parts.'

By the time Mr Ruche left the library, his mind was swimming with circles and lines. He caught a taxi back to the rue Ravignan.

Out on the bay tree in the courtyard, Sidney was hopping from branch to branch cackling madly. Sitting at a table beneath, Perrette sipped a strawberry soda and tried hard not to laugh. Mr Ruche was furious. He almost gave up reading the notes he had made. The parrot hopped from the branch and perched on Max's shoulder. By the time Mr Ruche read, 'Thales wanted to make true statements about a whole class of objects, statements that would be true for an infinite number of objects', Jonathan could stand it no longer. 'That's really horrible, Mr Ruche. Surely there must be a circle somewhere in the world, hiding out and waiting to ambush your theorem and disprove it?'

'No, not one', said Mr Ruche.

'Are you deaf?' said Lea. 'He said *all* circles. No exceptions.'

'It's a bit much', said Jonathan. 'I mean, it's practically fascist!'

Mr Ruche didn't answer. He was quietly impressed by their adolescent indignation. He enjoyed seeing them rail against conformity. It reminded him of his long conversations with Grosrouvre in smoky cafés by the Sorbonne.

'You can't escape from a theorem if it applies to you!' stated Lea, who looked like a cobra poised to strike.

Perrette glanced at her daughter, stunned to see her so vehement. She poured some soda water into her glass and added a drop or two of strawberry syrup.

'Don't you think your maths sounds a bit like Fate in tragedy, Mr Ruche?' she suggested quietly.

'*My* maths?' He was angry, but Grosrouvre would be pleased – he had won his bet.

Perrette elaborated: 'Don't you think there might be a link between maths and tragedy? I mean, they both originated in ancient Greece at about the same time, didn't they?'

Mr Ruche stared at her, stunned. This was something which had never occurred to him. Tragedy and Maths, Aeschylus, Euripides and Sophocles. That might be something worth studying. To Jonathan he said, 'Don't worry, theorems apply only to ideal objects.'

'In that case,' said Lea, 'he's got nothing to worry about.'

'Indeed not', agreed Mr Ruche. 'Theorems don't apply to people.'

'What about parrots?' asked Max.

'Not to parrots either.'

By dawn it was already unbearably hot, and the temperature continued to rise all morning. The only way to survive was somewhere air-conditioned. Jon-and-Lea headed off to the cinema on the Place Clichy. They preferred the big old cinema to the tiny screens of the ubiquitous multiplexes. This was a real cinema with comfortable seats, thick carpets and curtains that seemed to take a week to open. The screen was the size of a ship's sail. At the interval, they gorged themselves on ice cream.

Through some synchronicity of programming, *Land of the Pharaohs* was showing, which told the story of the building of the pyramids. They loved it and were sorry to leave the cool of the cinema. They walked back towards Montmartre, across the bridge at Caulaincourt which, unlike other bridges, spans a cemetery, so that the people crossing it are walking over graves. Still, it was better than a tunnel.

'There's not even a tree up here we can shelter under, but there are plenty down there in the cemetery', grumbled Lea. 'Same old story – the people who don't need anything get everything.' Jonathan watched her walk on ahead. She looked like a long-legged crow, Jonathan thought affectionately, and gave her a dig in the ribs.

'Don't touch me,' she screamed.

'Stop!' said Jonathan. 'I can smell your socks.' This was the code-phrase he used when he thought his sister was railing against the world.

Max was waiting for them in front of the bookshop, waving them towards the old artist's studio. It was unrecognizable: the floor was carpeted as thickly as the cinema in the Place Clichy. Sidney was perched on a high stool upholstered in purple velvet. Standing at the back, Mr Ruche smiled at them. Max had them sit on rush mats on the floor. There was a long silence and then the distant sound of waves. That was the signal. Sidney, in his raucous voice, began:

'Leaning against the stern, Thales watched as Ionia disappeared and Miletus, his home town, shrank to a dot on the horizon. Thales was heading for Egypt.' Sidney had the grave voice of a politician. He puffed out his chest and his eyes sparkled, and he stood tall to give himself more authority. 'Carried by the summer winds which blow only in periods of drought, the boat made the journey to Egypt, sailing into Lake Mariotis, where Thales boarded a barge to journey up the Nile.'

Sidney fell silent. Max stroked him gently and offered him a five-star snack of peanuts, almonds, hazelnuts and cashews. Mr Ruche took up the thread:

'After many days, stopping at small towns along the river, Thales saw the great pyramid of Cheops rising up from a vast plain. He had never seen anything like this in his life. The pyramids of Kephren and Mykerinos rose up on either side, but they seemed small by comparison. it was more impressive than anything he had imagined. He disembarked and, as he approached the pyramid, his pace slowed the closer he got. Overwhelmed, he had to sit down. A fellah sat beside him. "This pyramid that so impresses you...do you know", he asked, "how many died so that it might be built."

'"Thousands, probably", replied Thales.

'"Try tens of thousands."

'"Tens of thousands?"

'"Try hundreds of thousands."

'"Hundreds of thousands!" Thales looked at the man in disbelief.

'"Maybe more", said the fellah. "Why should so many die, to dig a canal, or build a bridge, to dam a river, build a palace, or raise a temple to God's glory? No. The great pyramid was conceived by the Pharaoh Cheops to remind people of how small they are in the vastness of things. The monument had to exceed our imagination in order to crush us: the greater the pyramid, the smaller humanity would seem beside it. He succeeded. I watched you as you approached, I saw the wonder on your face at such immensity. The pharaoh's architects wanted every man to admit that between this great monument and the humble man there was no common measure."

'Thales had already heard theories of this kind about the pyramid, but rarely so clearly or so starkly put. "No common measure!" The immensity of the monument was a challenge. For two thousand years, this edifice built by men remained beyond their understanding. Whatever the pharaoh had intended, one thing was certain: it would be impossible to measure the height of the pyramid. It was the largest construction in the known world and yet no one could measure its height. Thales decided to take up the challenge.

'The fellah talked all night, but what he told Thales no one knows. As the sun appeared over the horizon, Thales stood and looked at his shadow stretching out towards the west. However small an object is, he thought, sunlight can always make it seem larger. He stood for a long time, watching as the sun rose higher and his shadow grew shorter.

'"If I cannot measure the pyramid with my hands, then I shall do it with my mind", he thought. He stared at the pyramid for a long time. He needed an ally. He looked at himself, then at his shadow, and back again, and then at the pyramid. Finally he looked up at the sun – here at last was his ally. Whether it be the Greek god Helios, or the Egyptian Ra, the sun is changeless: it treats all things equally. This is a concept that the Greeks would later adapt to government and call democracy. In treating them equally, the sun held out the possibility that a common unit of measure might be found between the man and the monument.

'Thales considered the idea: my shadow relates to me exactly as the shadow of the pyramid relates to the pyramid itself. From this, he deduced that, at the moment his shadow was equal to his height, the shadow of the pyramid would be equal to its height. He had found a solution. Now he simply needed to put it into practice.

'He would not be able to do it alone, so the fellah agreed to help him. At dawn the next day, the fellah went and sat in the shadow of the pyramid. Thales drew a circle in the sand near the pyramid, the radius of which was his height, and stood in the centre, keeping his eyes fixed on his shadow. The moment the tip of his shadow touched the circumference, the moment it was equal to his height, he called to the fellah who immediately planted a stake at the point of the pyramid's shadow. Thales ran towards the stake. Using a rope held taut, they measured the distance from the stake to the base of the pyramid. Once they had calculated the length of the shadow, they could begin to calculate the height of the pyramid.

'The sand whipped about their feet, and a southerly wind began to blow. The Egyptian and the Greek walked together to the banks of the Nile where a barge had just docked. Thales boarded the barge. On the banks, the fellah smiled as the barge pulled away.

'Thales was proud of himself. With the help of the fellah, he had devised a trick. If he couldn't measure the vertical, he would measure the horizontal; if the height could not be measured, the shadow it threw on the ground could. Here was a way to measure the "great" with the "small", the inaccessible with the accessible; a way of measuring what was distant with what was near.

'Mathematics', Mr Ruche concluded, exhausted, 'is simply a series of clever tricks devised by great minds'.

This last phrase he said as much to himself as to his audience. On his purple-cushioned stool, Sidney was quite still; he looked as if he might be asleep.

'That's quite a story,' Perrette said, 'something of a Hollywood epic, in fact.'

'Thank you, my dear, that's quite a compliment. I'm very fond of Cecil B. de Mille.'

'The sound was good,' commented Lea, 'but it wasn't very visual. Still, it's a nice myth.'

'It's no myth!' Mr Ruche protested. 'Thales was a real person, the village Miletus really existed, the sun still shines and the pyramids still stand, the winds blow in summer and the Nile still flows.' He stopped for a moment. 'Even if it were a myth, what have you got against myths? This story was told by Plutarch, so it probably is true, but even if it weren't, Thales' theorem still holds.'

'Theorem? You didn't say anything about a theorem.'

Max smiled knowingly. He had said exactly the same thing when he and Mr Ruche had been rehearsing that afternoon.

All of a sudden a curtain came down, covering the windows and plunging the room into darkness. A white sheet descended on the far wall. Max started up an overhead projector and the motor hummed. Tiny lamps flickered on, lighting up the darkness. A blurred image appeared on the screen, and gradually came into focus:

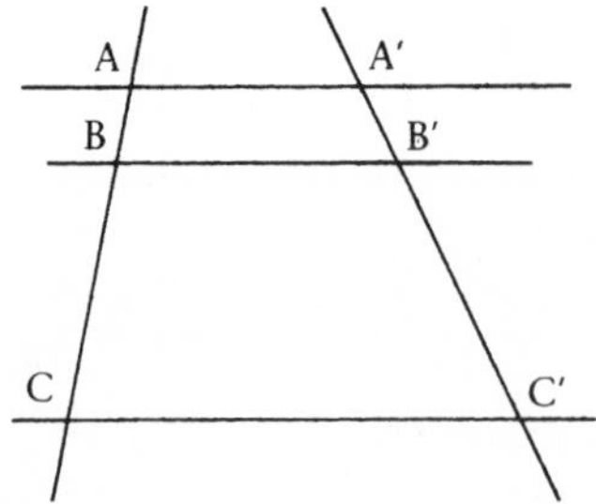

'Does this look like a theorem to you?' Mr Ruche demanded, triumphant.

'Yep!' Jonathan agreed.

'Next!' said Mr Ruche.

Max triggered the next slide:

$$\frac{\overline{AB}}{\overline{A'B'}} = \frac{\overline{AC}}{\overline{A'C'}}$$

'Aaaw!' the twins looked miserable. 'Doesn't look much like a Hollywood epic now, does it? More like some art-house movie. Too abstract...'

'Attention! Attention!' a harsh metallic voice interrupted. 'This is a theorem!' It wasn't Sidney. A light came on. By the window, a loudspeaker had been mounted near the ceiling. It was an old loudspeaker, the kind you saw in prison camps in old war movies. Max had brought it back from the market, and they had hooked up a tape recorder to it. Through the static it spat out: 'This is a theorem. Repeat, this is a theorem. Take three parallel lines, AA', BB' and CC', cut by two secants* into segments. The segments are proportional: AB-bar over A'B'-bar equals AC-bar over A'C'-bar.'

Jonathan and Lea were speechless. Only Sidney seemed unimpressed, perhaps jealous that this non-human was capable of speech. Max pressed 'Stop' on the tape recorder and all was silent.

'Not bad. Not bad for a beginner', Jon-and-Lea smiled at Mr Ruche, impressed by his first lesson as a teacher.

'You're quite right. This theorem is just the beginning of something which is one of the great achievements of Greek mathematics: the science of proportions. It is called "Thales' theorem" or the "theorem of proportions". Earlier, I mentioned that Thales had realized that the sun treats all objects equally. He was dealing with similarity, and the basis for similarity is *form*. Objects which look similar to each other have the same form. If proportions are conserved, then the form is conserved – in fact, you might more

*A secant in this context is simply a line that cuts other lines into two or more parts (segments).

accurately say that form is what is conserved when dimensions are changed but proportions maintained.'

He stopped for a moment to see the effect of his little speech. Jon-and-Lea were listening, really listening. A red dot appeared on the screen and danced around the formula like a fly.

'What we need to do is make the formula speak for itself', he enthused.

He had just remembered what Grosrouvre had always told him when they were studying for maths tests: 'You have to get the formula to talk. If you want to know what's going on, you have to question it.' At the time, Mr Ruche didn't really know what he meant.

'What was I saying?'

'You said something about making them talk?' said Jonathan.

'Yes, yes... "make the formula speak for itself"...so, what is Thales' theorem saying?'

'AB over A'B' equals AC over A'C' with a lot of bars all over the place', Lea replied.

'No, no, no. What does it *mean*? With language, when we say something, we do so to communicate an idea – well, usually. Maths is just the same. Thales' equation means' – the red dot settled on AB – 'that AB is to A'B' and AC is to A'C'.'

'What Thales' equation is telling us', continued Mr Ruche, 'is that the first pair and the second pair are in proportion to one another. That's the important word: proportion. This little theorem applies to all sorts of things: scale models, maps and blueprints, enlargements and reductions.'

Mr Ruche signalled to Max, who went to the far end of the room where, hidden in the darkness, was a photocopier. He quickly drew a parrot on a blank piece of paper, put it in the photocopier and pressed the button marked '50%', and gave the original and the photocopy to Mr Ruche.

'Reduction', said Mr Ruche. 'Same form, but smaller – in this case, the parrot is half the size.'

Max took the original and made another copy, this time pressing '150%', and again gave Mr Ruche the original with the photocopy. 'Enlargement', continued Mr Ruche. 'Same form, but larger – one and a half times larger than the original.'

Jonathan got up quietly, and grabbed the enlarged and reduced photocopies and held them up, mimicking Mr Ruche's voice: 'Same form, different size. Here,' he pointed at Lea, 'how much bigger is the enlarged parrot than the reduced parrot?'

Caught on the hop, Lea stammered, 'I won't say anything without my brief.' Sidney ruffled his feathers. Lea went on, changing the subject quickly: 'None of that tells us what Thales really did. After all, he was trying to measure a real pyramid, not something on a piece of paper.'

'I think you mean papyrus,' Jonathan chipped in.

'The equation is still the same whether it's written on papyrus or paper.'

For a moment Max wondered what it would be like if it did change things – if a plus sign on paper became a minus sign on parchment, or a multiplication symbol on vellum.

'How many times bigger is the parrot?' Jonathan wouldn't let it drop.

No one bothered to answer. The equation vanished from the screen and the next slide appeared:

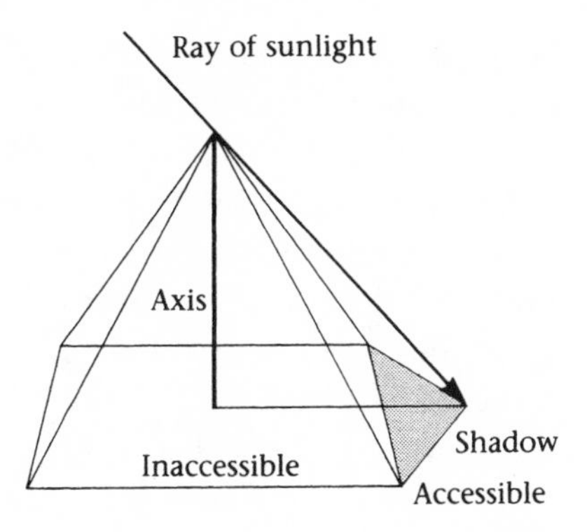

Mr Ruche went on: 'If it had been a tree, say, or the obelisk in the Place de la Concorde – which was in Egypt before it was brought to Paris – Thales' job would have been easier. He could simply have measured the shadow at the right moment. But a pyramid's base is of course wider than its summit. The base of the Great Pyramid of Cheops is square, and the axis rises from a point exactly in the centre of the square. The height of the pyramid is the height of the axis. But all that Thales could do was measure the length of the shadow of the axis. And he couldn't even measure the whole shadow – only the part that extended from the base. The part within the pyramid was obviously inaccessible.'

'So it was all for nothing, then.' Lea sounded indignant.

'That's what I thought at first. But after a while I found the answer. Thales would have had to take his measurements when the rays of the sun were exactly perpendicular to the side of the base.'

'Meaning?' Lea asked.

'Hold on, let me remember. If the rays were perpendicular to the base, that would mean that the hidden part was exactly half the length of one side.'

'Well, I don't understand', Lea said.

'Me neither', said Jonathan.

'Dinner time!' shouted Perrette.

'Saved by the bell!' thought Mr Ruche. 'Well, I could eat a horse.'

The next day, Jon-and-Lea had a half-day from school. The moment they came in Mr Ruche said, 'Hurry up, Albert is coming over.'

The doorbell rang. Albert was a young sixty-year-old. He wore a dirty grey cap, bottle-end glasses and had a stubbed-out cigarette between his lips. 'Afternoon, all!' He grabbed Mr Ruche, wheelchair and all. He knew exactly what he was doing. Since the accident, Albert had been the bookshop chauffeur and he took Mr Ruche everywhere in his Renault 404 convertible. It was he who had been ferrying Mr Ruche to and from the library.

Mr Ruche always called Albert 'one of a kind'. Mr Ruche himself was one of a kind in his way. Albert had always refused to use a radio cab. He couldn't understand how customers could bear to travel in a car with the radio constantly going. 'car 57...car 57... pick-up from the Boulevard Vaugirard'. Albert worked from the railway stations and sometimes picked up customers who flagged him down in the street. He also had a number of private customers like Mr Ruche. The two of them had become great friends after the accident, and whenever Albert took a day off he would call round early and pick up Mr Ruche, and the two would drive into the country for the day with a hamper full of food on the back seat.

Max had class that afternoon, but Perrette gave him permission to go with the others. Everyone climbed into the 404 – even Sidney – and Perrette watched from the doorway of the bookshop as they drove off. Mr Ruche had refused to tell them where they were headed. They drove down the Avenue de l'Opéra, past the Palais-Royal and through the Cour du Carrousel. Albert braked hard and parked beside the pavement. There in the great courtyard called the Cours Napoléon stood the Glass Pyramid. They all clambered out into the courtyard.

'The Great Pyramid of Cheops is 4639 years older than the pyramid you see here. One is stone, the other glass; one is on the banks of the Nile, this one is on the banks of the Seine.' As he spoke, Mr Ruche took out a sketch pad and some pencils. 'Thales' idea that the sun treats all objects equally is based on the fact that the sun's rays are parallel to one another. Though not precisely true, the sun is far enough away for the estimation to be correct. Here is the situation when Thales made his measurements.'

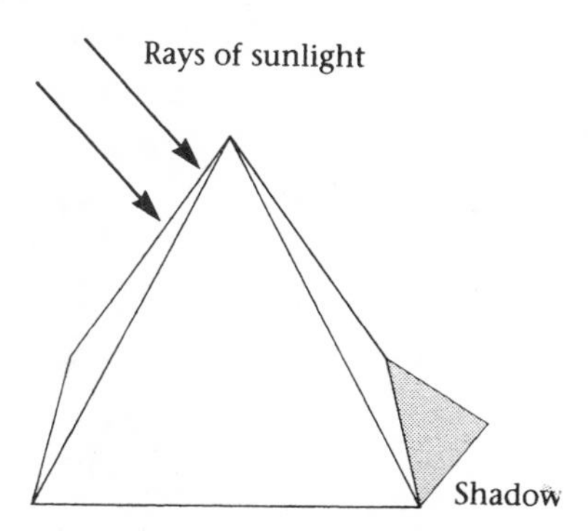

As soon as Mr Ruche started to draw, Sidney hopped onto the old man's shoulder as though he wanted to see better. 'Since the pyramid Thales was measuring wasn't transparent, like this one, I'll explain everything. First I'll take away the things that are stopping him from seeing the inside and I'll trace the axis of the pyramid. The height of the pyramid is the length of the axis – that's what Thales is trying to measure.' Mr Ruche rubbed out the shaded sides and drew an axis.

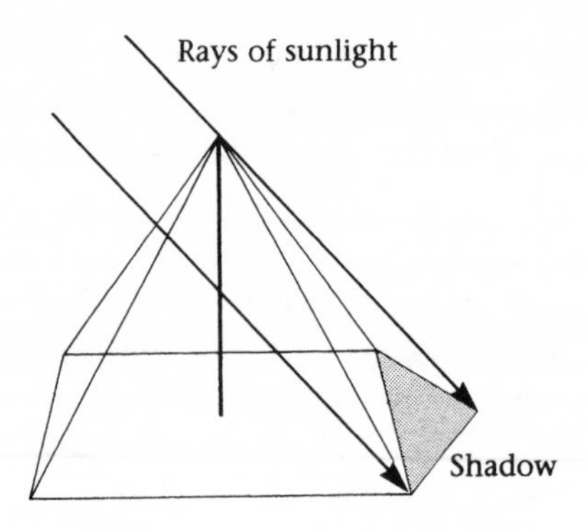

'Right. Let's move on.'

Mr Ruche was flailing wildly now, so Sidney hopped off his shoulder and onto Max's. Mr Ruche, meanwhile, rubbed out the sides of the pyramid altogether, then traced a horizontal line from the base of the axis to the point of the triangle which represented the shadow.

'If the pyramid were transparent, this is how the shadow of the axis would appear. This is what Thales was trying to measure.'

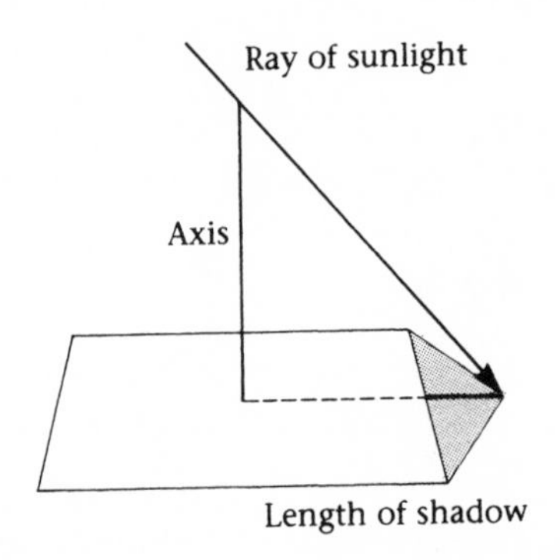

'The dotted line is the part of the shadow that falls *inside* the pyramid; Thales can't measure it, because he can't get to it; the fat black line from the base to the point of the shadow *can* be measured. In fact, it's the *only* thing he can measure.'

Mr Ruche rubbed out the shadow. He extended the axis, then wrote the letter A at the base of the axis, the letter H where the line cut the wall of the pyramid and M where the point of the shadow had been. He put the first drawing and the last drawing side by side. 'There we go: Before and After, just like in the slimming ads.'

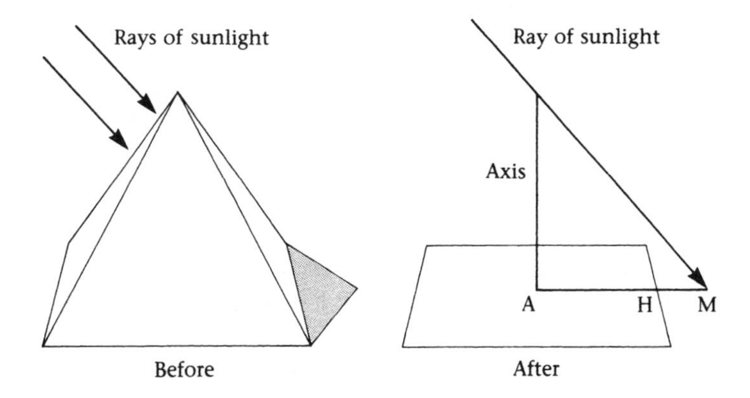

'We've stripped everything down to bare essentials. Forget the mass of the pyramid – rub it out. All we need to keep is what relates directly to the question. Erase, simplify, purify: that's what Thales did. I think that's how most mathematicians work. They call it "abstracting". For a mathematician, this would be enough to solve the problem.'

'What?!' Jon-and-Lea protested.

'If Thales had been measuring an obelisk, this is all he would need. He would simply measure the length AM along the ground, but he was measuring a pyramid and the first part, AH, was inside the pyramid.'

'Then he's stuffed,' Jon-and-Lea shouted triumphantly.

Mr Ruche ignored them. He could see that some tourists had stopped nearby and were following the proceedings.

'All right. Most of the time, the sun's rays fell at any old angle on the pyramid of Cheops and it cast any old shadow on the sands. When that was the case, Thales couldn't do anything.' Mr Ruche drew another diagram:

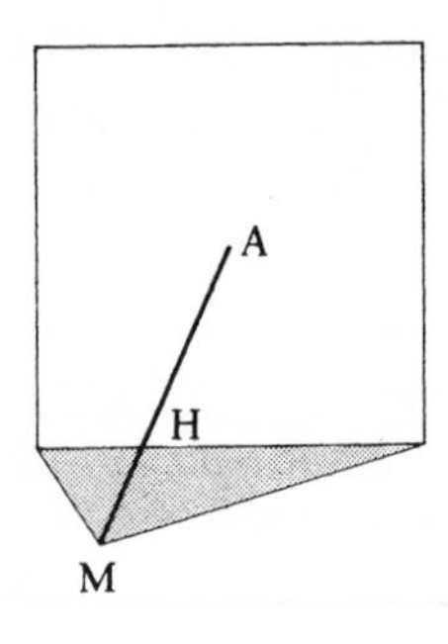

'Don't forget, maths is all about finding a clever ruse. Thales must find a particular case that will help him out. He finds it by deciding when to measure the shadow – the precise moment of the precise day when the sun's rays will be perpendicular to the side of the base. I mentioned this at the house, but you didn't seem to understand. What Thales could not measure directly, he would deduce by reason. What did he have at his disposal? The only part of the pyramid to which he had access is the side, so he will use that.'

Mr Ruche quickly sketched a new diagram:

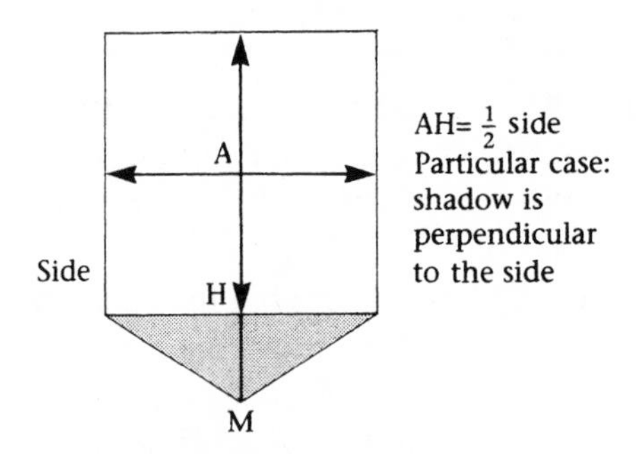

Satisfied, he looked round at his audience. There were now more tourists clustered round. As he was closing the notebook, Jonathan asked:

'How did Thales know when the shadow was perpendicular to the side?'

'There's the catch', thought Mr Ruche, and shot him a glance.

'Good question, and one that I asked myself.' Reluctantly, he reopened his notebook.

'Thales didn't have a protractor to measure the angle, he had something better – the orientation of the pyramid itself. The architects had designed the monument so that one side would face due south. The shadow would be perpendicular when the sun is at its highest: at noon precisely.'

'Just when it's at its hottest', said Jonathan.

'No pain, no gain', said Lea philosophically. 'I don't suppose history tells us if Thales got sunburnt, going into the desert at midday?'

'Midday, yes, but he was in the shade', said Mr Ruche. 'Might I remind you that Thales was measuring the shadow. To measure it, there had to be a shadow and in that case he could stay in the shade.'

A laugh rippled through the crowd.

'What about this shadow, Mr Ruche? Did the pyramid cast a shadow every day at midday?' asked Max.

'No.'

Jonathan interrupted: 'In any case, the shadow would have to be visible, which means it would have to extend outside the pyramid – at least that's what I understood.'

'And it has to extend past the pyramid at midday, because at any other time it's no use to Thales anyway', added Lea.

'And the shadow has to be exactly as long as the pyramid is high – that's an awful lot of ifs and buts', said Jonathan.

Mr Ruche waited until they had finished.

'You're right – the pyramid does not cast a visible shadow every day at midday. That's the problem. For there to be a visible shadow, the sun cannot be too high at its highest point.

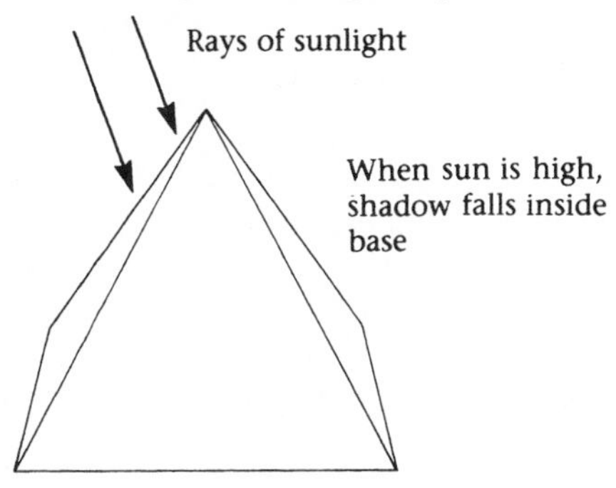

'To recap: the shadow needs to be equal to the height of the pyramid and it must be perpendicular to the side. This is not a matter of abstract geometry, but involves astronomy, geodesy* and geography, so we need to think back. The Great Pyramid of Cheops is at Giza, at 30°E in the northern hemisphere. It's considerably south of Paris, just above the tropics. For the shadow to be equal to the height of the pyramid, the sun's rays must be at 45°. At Giza at midday in the summer the sun's rays are almost vertical, so for a great part of the year the pyramid will cast no shadow at all. Another thing: for the shadow to be perpendicular to the base, it has to fall north–south. There are only two days in the year when all these conditions are true: 21 November and 20 January. You see, Lea, it may be the middle of the day, but it also has to be the middle of winter – if Thales was going to catch anything, he would have been more likely to catch a cold.'

A group of Japanese tourists pushed forward. One wanted to buy one of Mr Ruche's sketches, another took a photo.

'The theorem may be true in general, but the measurements seem to be pretty specific. So what was it? I mean, he did measure the height of the pyramid, didn't he?' Lea asked.

'The only thing he had to measure length with was a piece of rope. He needed a unit of measurement, so he used the *Thales* – he measured it in units of his own height. Measuring the shadow, he got 18 Thales. He then measured the base to be 134 Thales, and divided by 2 to get 67 Thales; adding them together, he discovered that the Great Pyramid of Cheops was 85 Thales high. A Thales, he later worked out, was equal to 3.25 Egyptian cubits, making 276.25 cubits or 482 feet.'

Mr Ruche didn't mention how long it had taken him the night before to work through the calculations – or how many times he had to give up and start again.

'This pyramid', he pointed to the glass Pyramid du Louvre, 'is...'

He fumbled with his notebook, when suddenly Albert piped up:

'71 feet high and 113 feet long on each side.'

*Geodesy: a branch of applied mathematics concerned with the determination of the size and shape of the earth and the exact positions of points on its surface, and with the description of variations of its gravitational field.

Everyone looked at him in surprise, and he nervously touched his cap. 'I hear it every time I drop off a tourist', he added apologetically.

'To answer any other queries, I've prepared a couple of drawings.' Mr Ruche tore out the pages and handed them round.

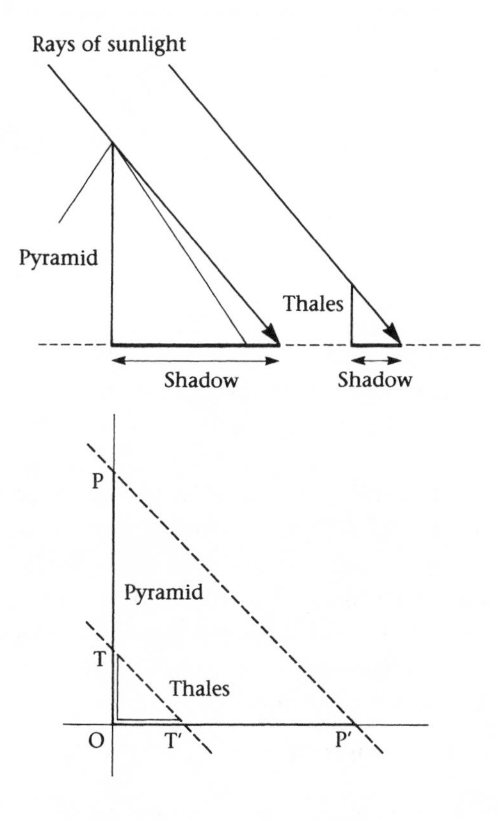

'The relationship between the heights and lengths we can write as:

$$\frac{\overline{OT}}{\overline{OP}} = \frac{\overline{OT'}}{\overline{OP'}}$$

'And here', Mr Ruche said, 'is the drawing I showed you at the slide show last night, which represents the theorem of Thales.'

He handed over the last diagram. The abstraction of the problem had worked marvellously well. Here was the theorem in its pure form, without ornament or unnecessary baggage. This was a real mathematical diagram.

'The theorem', concluded Mr Ruche, 'explains what takes place when two secants cut parallel lines.'

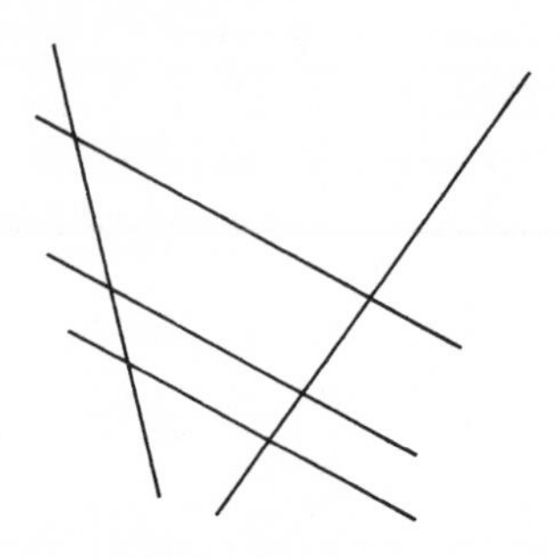

A huge round of applause greeted Mr Ruche's concluding sentence. The name of Thales rang out in all manner of accents, such as '*telis*' and '*tahlis*', and one American shouted enthusiastically, 'Yeah, Taylis!' The Japanese tourists were so impressed that they wanted to give money. Some months later, a photo appeared in the arts supplement of a Tokyo newspaper. It showed Mr Ruche enthroned on his wheelchair with Max on one side – Sidney perched on his shoulder – and Albert, who had instinctively removed his cap, but kept his cigarette-end. In the background of the photo was the famous Pyramid du Louvre. Underneath the photo was the following caption:

高齢のフランス人学者は、建築家イェオ・ミン・ペイの設計によるルーヴル美術館のガラス製ピラミッドの高さを、古代ギリシアの数学者タレスの影を使う方式で測定する。

The sun had gone down behind the walls of the Tuileries gardens, and it was beginning to get cold. Rather than heading directly north, Albert drove along the Seine and into the Place de la Concorde at precisely the moment the streetlights came on. They drove twice around the square so that everyone could get a good look at the Egyptian obelisk, then, taking the rue Saint-Honoré, Albert took them past the great column on the Place Vendôme.

'As you can see,' Mr Ruche said in a rather tired voice, 'you can take an obelisk anywhere, but it's a little more difficult to move a pyramid.'

'Not to mention measuring one', said Max.

'That's always the way', said Mr Ruche. 'My old maths teacher used to say, "Then all you need to do is apply the theorem" and put down the chalk. Easy for him to say...'

'I think maths is easy', said Lea. 'It's applying the principles that's hard.'

'I'd say that maths is hard and applying the principles is even harder', countered Jonathan.

'I think you're being a bit dramatic', said Mr Ruche. 'Thales' theorem is true regardless of what it is applied to. To demonstrate it, however, Thales used a very specific example – measuring the pyramid. He chose the moment when the ratio between the pyramid and its shadow was 1:1, because that was the simplest solution.'

'It might have been simplest, but it was also a pretty rare one', said Jonathan.

'That's normal – a particular set of circumstances will always be rarer than a general one. Life is like that: you have to choose between what is complicated and happens often and what is simple but happens rarely', Mr Ruche said philosophically.

'I'd go for simple and often', said Jon-and-Lea together.

Max piped up: 'Mr Ruche, when you were telling the story back at the house, you said that Thales left Miletus in the middle of a heatwave and that he didn't stop until he got to Cheops. But a minute ago, you said he measured the pyramid in the middle of winter. Surely it didn't take him six months to get there?'

Mr Ruche came down to earth with a bump: he'd been caught out.

'Maybe he did stop along the way – I don't know. Perhaps he stopped off in Alexandria. No, what am I saying? Alexandria was built much later. Thebes, then. Actually, I think Thales probably set up camp beside the pyramid and waited for the right moment to measure it.'

'And what about the fellah? The one who helped Thales? What happened to him?' asked Max.

Mr Ruche nodded – he had completely forgotten the fellah.

'Yes, without the fellah he couldn't have measured the pyramid', chorused Jon-and-Lea.

'You're quite right. Without the fellah, Thales would never have been able to measure his own shadow and at the same time mark the point of the pyramid's shadow. Thales' theorem needs two people to be able to prove it.'

'It should be called Thales and the fellah's theorem', decided Lea. 'Credit where credit's due.'

Mr Ruche made a mental note to ask himself, 'Who is the fellah?' every time he came across a new theorem. As they drove up to Montmartre, he thought about the lessons he had learned in teaching Thales' theorem and how he should proceed with his history of mathematics. His stories should be as close to the historical truth as possible. The twins were proving to be excellent at cross-examining; he knew he would get nothing past them. This was going to be more difficult than he had thought, but more exciting too.

'Did you know that Thales once predicted an eclipse, Mr Ruche?' Jonathan asked suddenly.

'Yes.'

'You never said!'

'No.'

'From what I read,' Jonathan went on, 'it wasn't his theorem that made him famous at the time, it was the fact that the eclipse happened exactly when he predicted it would.'

Lea was a little shocked by this sudden display of wisdom, and gave Jonathan an evil look before turning to Mr Ruche. 'Thales' twit of a servant-girl should have kept her mouth shut. She was completely wrong when she said, "If you can't even see what's in front of you, what makes you think you can see what goes on in the heavens?"' she remarked acridly.

Albert braked suddenly and Lea's head bumped against the window, but she went on, 'What she meant was, "If you spend all your time looking at the sky, you're bound to fall in a hole in the road."' And without waiting for Mr Ruche to answer, she told Albert to stop the car. She climbed out and Jonathan followed her.

'Why didn't you tell me the story of the eclipse first? Are you going solo now?'

'You could find things out by yourself if you wanted to. Just remember that two is also two times one.'

They continued along the central reservation, Jonathan walking ahead angrily. 'I don't have to tell her every little thing I do. She has to get it into her head that we have to live our own lives as well as being twins', he said to himself. Then he thought about the eclipse. By studying the sky, Thales was able to calculate when the eclipse would happen and so free himself of the fear of being suddenly plunged into darkness.

'What you said about falling in the hole', he said to his sister. 'I think Thales accepted the risk that he might fall and find himself suddenly in darkness.'

'A little local darkness', added Lea.

'Exactly', said Jonathan. 'Thales accepted the risk, for by studying the sky he could predict the darkness that was going to fall on everyone and scare them half to death.'

Lea stared at Jonathan, surprised. It was unlike him to say so much, or to explain himself so completely. Maybe Perrette's sudden revelation about their conception had upset him more than she realized. She walked beside him and thought how glad she was that they were born twins, so that they could face this thing together. 'Two is also one plus one', she thought. She stopped and rubbed the lump on her head from Albert's clumsy braking, then grabbed Jonathan's arm. 'Falling in the hole was the price he paid to be free of his fear of the future, is that what you mean?'

They agreed that, all things considered, Thales was a good guy. Jon-and-Lea decided to adopt this man who had tamed the shadows and the darkness in the world.

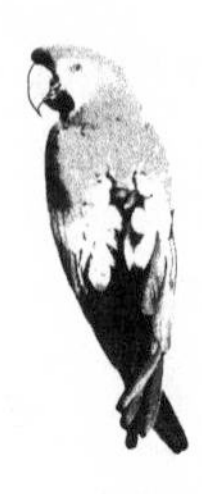

CHAPTER 4

The Rainforest Library

Books do furnish a room; an outline of the history of mathematics & a parrot without a passport

Two weeks later an articulated lorry pulled up outside the bookshop. The windows rattled like fireworks exploding on Bastille Day. Someone knocked on the door of the garage room and Mr Ruche opened it. A little guy stood there, waving a docket:

'Got a street name here, but no number. You Mr Rich?'

'Ruche', corrected Mr Ruche, who looked out and saw one of the men opening the doors of the lorry: it was full of crates. He couldn't believe it – Grosrouvre's library had arrived!

'You listening, or what?' the little guy shouted right into Mr Ruche's ear. 'I said you was lucky to get this stuff at all. The ship they was on nearly sank halfway across the Atlantic. One of the sailors told me that just when the captain said to chuck the cargo in the sea, some Cuban warship showed up and helped 'em. Could say it's a miracle any of this stuff got here at all. 'Don't believe in miracles myself. If they didn't get thrown over then they was supposed to get here.'

As he talked, the removal men emptied the lorry and the crates piled up in the new library.

'Books, nothing's heavier than books', grumbled one of the men as he staggered past. 'And folk always pack 'em right up to the top. You can tell it's not them that's going to have to carry 'em!' He sat down, mopping his face. 'All the way from Brazil, they come.' He pointed at a ticket on the crate. 'Usually it's wood that gets shipped over from Brazil. You see it on the docks, huge great lumps of wood. I tell you, the trees round 'ere are like matches compared to that lot.'

'They didn't get wet, I hope?' asked Mr Ruche bluntly.

'I dunno, do I? We just move 'em.'

Mr Ruche started inspecting each of the crates carefully. He wheeled himself easily round the boxes, running his hand along the wood, but there was not a trace of damp.

The delivery men left. Mr Ruche heard the lorry driving off down the rue Ravignan; then, silence. The studio Mr Ruche had had converted into a library was lined from floor to ceiling with the new bookshelves built by the carpenter from the rue des Trois-Frères. A flood of cold light fell from the skylight. Like all artist's studios, it was north-facing. Soon, the shelves would be filled with books. They'll be fine here, he thought, no damp and not too much sunlight.

Perrette slipped a crowbar under the top of a crate. The wood creaked and made a noise like a nut being cracked. Mr Ruche turned just in time to see the top come off. Books filled the crate to the brim, stacked flat one on top of one another. 'The fool!' shouted Mr Ruche, 'the books at the bottom will be completely squashed.' Perrette took one out and studied it carefully, then looked over at Mr Ruche, incredulous. She was holding a sixteenth-century book in her hands and it was in perfect condition. She felt strangely moved. She handed it to Mr Ruche, but he wouldn't take it, so she put it on the nearest shelf. The first book!

Mr Ruche watched Perrette carefully. Lids popped like cracked walnuts as she opened the other crates. The squeaking of the wheelchair broke the silence. Mr Ruche wheeled himself towards the shelves. With meticulous attention, he studied the books that Perrette had already put there. He didn't touch them, he was happy simply to look at them lovingly and read the titles on the spines where he could. This was a tiny fraction of Grosrouvre's library – the rest was in the crates.

'He must have been rich to be able to afford all these,' Mr Ruche said to no one in particular. 'Must have been?' Perrette was surprised. 'You think he isn't any more? You don't think he's bankrupt – or dead?'

'Of course not. What makes you say that? I'm sure he'll be sending us more books.'

Perrette didn't look convinced. She made no move to go to the bookshop, though it was due to open any moment now. She wanted to stay here, in the library, with the books and Mr Ruche. He sensed it himself, and so decided to go with her to the shop. It was the first time since his accident that he had gone back into the bookshop.

Their first customer was an elegant young woman with chiselled features. She browsed through the new books, before buying a copy of *Skin Deep*, the bestseller about plastic surgery by Dr Larrey.

Perrette walked back over to Mr Ruche: 'I didn't see any labels on the crates.'

'There weren't any.'

'It would have made our job a lot easier.'

'In his letter, Grosrouvre told me he hadn't had time to sort out the books crate by crate.' He stopped. 'Did you say "Our job"?'

Perrette blushed. 'I'll help you sort them out – if that's all right with you?'

'Of course. I didn't want to ask – you have so much work with the shop. It will be just like when we started working together.'

'Are you going to keep them?'

'Keep what?'

'The books, of course.'

'I'll keep them safe, at least until Grosrouvre tells me what he wants me to do with them.'

'He's a strange man, your friend, don't you think? What could have been so urgent that he didn't have time to sort out the books when he was boxing them up?'

'I've been wondering that myself. And it's not the only thing I've been thinking about. Why has he sent his whole library to me?' I could have been dead for years, for all he knew. He'd have got his letter back marked "no longer at this address". In fact it could easily have happened, he wrote "1001 Pages" instead of "A Thousand and One Pages".' A sly smile spread across his face. 'Maybe I should send the whole lot back to him.' Mr Ruche imagined Grosrouvre's face if an enormous delivery of crates marked 'return to sender' arrived back at his house in the middle of the rainforest.

'So you know his address?' Perrette asked innocently.

Mr Ruche was speechless. His revenge fantasy had been short-lived; he didn't know Grosrouvre's address, didn't even have a telephone number for him. He had no way of contacting his friend. It was as though Grosrouvre knew that the communication would be one-way. Perrette picked up the telephone and dialled international directory enquiries. the international operator was categorical: there was no Elgar Grosrouvre listed in Manaus, Brazil. Mr Ruche remembered that in his letter Grosrouvre had said he lived *near* Manaus, but not exactly where. 'In a country like Brazil, if you say "near", you could still be a hundred miles away', said Perrette, still holding the handset. 'Sorry?' she said to the operator, 'So you'd need the name of a village or a town?'

She hung up. Mr Ruche shrugged his shoulders. He was trapped. Grosrouvre had always been like this. He decided what he wanted and tried to get everyone else to fit in with his plans. It usually worked, and you found yourself doing exactly what he intended you to do.

'You're sure all this is really from your friend Grosrouvre?' asked Perrette.

'If it seems so to me and to all the world it is not therefore evidence that it is so; but what one might wisely ask is whether there is reason to doubt it?'

Perrette looked at him, astonished.

'Wittgenstein, Perrette, why would there be any reason to doubt it?'

A woman in her fifties pushed open the door and asked for a 'book about fishing or something like that'. It was a present, she explained, for her husband, who had just retired. Mr Ruche left Perrette to it. She would be better off giving him a fishing rod and some bait, thought Mr Ruche as he headed back to the library.

Mr Ruche plunged his hand into the nearest crate and, for a moment, he imagined the boxes lying thousands of metres down at the bottom of the ocean. He felt dizzy at the very idea of this, the most complete mathematical library in the world, scattered on the ocean floor.

He pulled himself out of this nightmare vision. A book – very solid and very real – was gripped in his hands. He stroked the fine grain of the leather binding and felt his fear drain away. He looked across at the empty shelves. The books were here and the shelves ready, waiting for them. Grosrouvre had entrusted his library to him and Mr Ruche would make sure that no harm came to it.

When Jonathan and Lea went into the library, they found Mr Ruche poring over the books. His eyes, so often glazed, were shining with the fervour of a man half his age. Mr Ruche seemed transformed by this library which had come from the ends of the earth. They christened it the Rainforest Library.

Mr Ruche was like a child trying to open all his presents at once. He wanted to take out all the books and put them on the shelves immediately, so that he could gauge how many there were. But it would be impossible to use a library where the books were placed in no particular order. He would be out of the frying pan and into the fire – not a good metaphor to use where books are concerned. He decided to be reasonable. He would have to devise a cataloguing system for the Rainforest Library.

When Mr Ruche had first opened his bookshop, he sorted the books into categories: fiction, non-fiction, thrillers, science fiction, travel books, a small poetry section, and even a foreign language section with novels for tourists who stopped in on their way to the Sacré Coeur. Over time, he remembered, the cataloguing system had changed and evolved.

Grosrouvre wasn't making things easy for him. 'If I could just get in touch with him I could ask him how to go about organizing the library. He could at least have sent me *his* inventory of the books. How am I supposed to decide how to catalogue books, when I don't even know what they all are? How can I sort out maths books if I don't know anything about maths? When I was twenty, I wasn't interested in maths, now he is forcing me to dive into his world at the age of eighty-four. He's a devious old bugger.'

His anger evaporated as quickly as it had appeared. Grosrouvre hadn't done this deliberately. Despite all the ironic quips, his letter had been serious. His old friend had sounded almost panicked. Perhaps something had forced Grosrouvre to ship the library to him, but what?

'The packages I had made up have all come apart. I've had to repack the books higgledy-piggledy into packing crates. You'll have to sort them out and catalogue them however you see fit, πR. It's not my problem any more!'

'No, no...it's my problem now!' grumbled Mr Ruche, 'and that's just how he would have wanted it.'

He decided to catalogue the books chronologically and by subject matter: each book would be classed according to the date of first publication and its subject matter. Firstly he had to divide the history of mathematics into historical sections; then divide it into the different subjects which would form the subsections. Since new disciplines evolved all the time, the subsections would not necessarily be the same for each historical section. Some mathematical concepts become redundant and disappear, some are absorbed into others, and sometimes a completely new idea appears. Setting out this cataloguing system would be like recreating the whole arc of mathematical history. Mr Ruche would have to study geography and history as well as maths to map it all out.

'Grosrouvre goes off to live in the middle of the jungle, but I'm the one doing the exploring, in my own backyard.'

He would accept the challenge. After thinking for a moment, he decided that there were three great periods in mathematical history (though he might refine these later):

Section 1: Maths in Ancient Greece (though the period was a little longer than the classical Greek world, from 700 BC to AD 700, perhaps).
Section 2: Maths in the Arab world (from AD 800 to 1400).
Section 3: Maths in Western civilization (from 1400 to the present).

The subsections would prove more difficult. You had to be able to answer one simple question: what is mathematics about? So, what *was* mathematics about?

Shapes, space, volumes and numbers. Those were the first things: *Geometry*, *Arithmetic*. The categories were a little rudimentary, he thought, but before consulting a dictionary or an encyclopedia he tried to remember the names of the classes he took when he was at school. More than sixty years after leaving, Mr Ruche managed to remember: *Geometry*, *Arithmetic*, *Algebra*, *Trigonometry*, *Probability*, *Statistics*, *Mechanics*. Geometry was to do with shapes; arithmetic with numbers; algebra with symbols; trigonometry with angles; probability with likelihood; statistics with large quantities of data; mechanics with bodies in motion.

Mr Ruche wanted the decision to be democratic and had invited Perrette, Max and the twins. Max arrived carrying an armful of things: drawing paper, an eraser the size of a tennis ball, a ruler and some coloured pencils – he hated markers. He Sellotaped several sheets of paper together and stuck them to the wall.

With his notepad open on his lap, Mr Ruche explained to them how he intended to arrange the books in the Rainforest Library. Geometry was accepted unanimously as a category. Max drew a square on the makeshift whiteboard he had made and wrote 'Geometry' in it. Arithmetic was a little more difficult. Some of them thought it should be included in Algebra. Mr Ruche tried to defend two separate sections:

'Arithmetic comes from the Greek word *arithmos*, meaning "number".'

'He never misses a chance to get his Latin or Greek in', thought Lea and asked, disingenuously, 'Where does the word "algebra" come from?'

Mr Ruche hadn't the faintest idea. He went back to his notes.

'Arithmetic is the science of whole, natural numbers: 1, 2, 3, and so on. Algebra deals with equations. So in arithmetic we study the nature of whole numbers and their properties: whether they are odd or even, divisible or prime. In algebra, we try to solve equations, but we're not concerned with the properties of the result. What is important is the form of the equation, which will determine the type of result we get.'

His audience looked sceptical, so he continued:

'"The sum of two whole even numbers is a whole even number". This is an arithmetical statement, but the equation $ax^2 + bx + c = 0$ is algebraic.'

He thought he detected a glimmer of understanding.

His final, decisive argument was that, while arithmetic began in Greece in the sixth century BC, algebra was not discovered until much later. Max drew a box for each section as Mr Ruche moved on.

'As the name suggests, trigonometry concerns the measurement of triangles, but it is concerned with the angles rather than the sides. It's sometimes called the science of shadows, if you get the reference.'

'Yeah, Taylis!' shouted Jonathan in his best American accent.

'It is the science of incline and the orientation of objects – anything, in fact, that can be measured by angles. Using the *sine* and the *cosine*, you can calculate an angle without having to measure it. The sine and cosine of an angle are numbers.'

This split his audience into two camps: those who wanted an autonomous category for trigonometry, and those who wanted it to be included in an existing category.

Jonathan suggested, 'Since the sine is a measurement of an angle and angles are in geometry, it should be part of geometry.'

Lea naturally, opted for making it part of arithmetic: 'Since sine and cosine are numbers, surely it should be part of arithmetic.'

But Mr Ruche won the debate: 'Since trigonometry is a marriage of both arithmetic and algebra, I think we need a separate room for the newly-weds.'

Max set about drawing another box.

Next came probability.

'The probability of Max finding a parrot being beaten up in a warehouse is practically nil,' suggested Lea, 'but he did and I for one am very happy that he did.'

If finding Sidney meant that probability had its own category, that was good enough for Max. Next came mechanics, which, Mr Ruche pointed out, was about theory rather than practical mechanics.

'Mechanics is about bodies in motion. What causes objects to move?' Mr Ruche answered his own question: 'Forces acting upon them. The mathematician tries to define a force as a formula.'

Silence. Mr Ruche wished that Albert was there: as a man who had much to do with cars, he would certainly have had an opinion. Max drew another box. Perrette asked why statistics didn't have a category, and Mr Ruche said he had decided that as statistics dealt with specifics, it was too empirical to be a subsection of maths.

'Do you know what you've forgotten?' asked Perrette. *'Logic!'*

'I haven't forgotten about it', said Mr Ruche confidently. 'Logic was devised by Aristotle who was, unless I'm very much mistaken, a philosopher, not a mathematician.'

'If there's no logic in math, I don't know where there is any!" said Perrette.

'Maybe, Perrette, in the ideas.'

'And in the reasoning used to work things out; no reasoning, no maths!'

'That's logical, Mum', said Max, drawing another box. Mr Ruche had been beaten hands down.

'What about new maths?' asked Max.

There was a roaring argument about this, with Perrette reminding everyone that 'new' was an adjective indicating that something was recent.

'Maybe it is an adjective, but new maths is all about sets, and a "set" isn't a number and it isn't an angle or an equation, a sine, a cosine or a probability, so...'

Perrette gave in, but only on condition that they wrote it as one word like a noun. Max drew the box and wrote 'Newmaths' in it. They looked at the whiteboard:

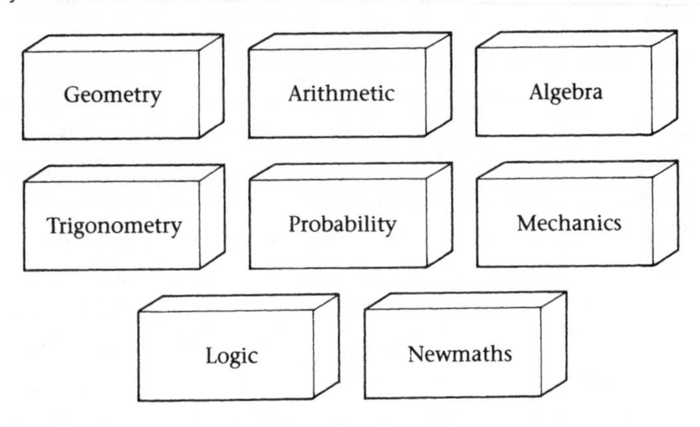

They counted up and found that there were three main sections, divided by time with eight possible subsections, so there would be twenty-four different categories in the Rainforest Library.

Max, with Sidney perched on his shoulder, walked slowly round the pet shop on the Quai de la Mégisserie. There were turtle-doves and racing pigeons, ducks, doves, tiny birds from Mozambique, birds with blue foreheads, red breasts, long tails. Canaries of all sorts, songbirds, parrots, cockatiels and cockatoos, toucans and macaws. A white dove with a coral-coloured beak and orange cheek-feathers. Rabbits of all sorts and sizes; gerbils and hamsters; guinea pigs and chinchillas. There were lizards, chameleons and a huge boa constrictor. Max stared at the snake for a long time and then moved on.

He had taken the precaution of wearing a hat to cover his red hair. After all, you could never tell. On a wall in a small side street he saw a small plaque which read 'Rue Jean Lantier. This street is named after a man who lived here in the 13th century'. Seven hundred years seemed a long time. Max had just found out that some animals lived to be more than a hundred. How old was Sidney, he wondered. He had come to try and find out.

All along the river between the Seine and Chatelet were hundreds of outdoor booksellers. Books lined the quays for miles and miles with barely enough room to walk between them and the road. There were pet shops here too, with every kind of bird in the world, except of course those which were endangered species. It was illegal to sell endangered species, but if you had the money...

Max went into one of the pet shops that specialized in birds. At the door there was a sign saying 'no animals allowed'. He burst out laughing. He wondered what part of the world Sidney came from, what species of parrot he was. Max looked at a poster illustrating hundreds of species of parrot, but couldn't find Sidney on it anywhere. A man in the shop was pointing out the different geographical regions where parrots lived to his companion – principally, Central Africa and South America – but parrots could also be found in Asia and India.

Max headed for the pet food section. There were two brands of parrot food, ordinary and superior. The superior brand had a large variety of nuts and grains: sunflower seeds, millet, rice, wheat, peanuts; while the ordinary variety seemed to be just sunflower seeds and millet. Max picked up a large bag of the superior mixture and a handful of parrot treats with honey. Sidney was really excited. Max stopped in front of a list of all the pet shops in Paris. Beside it was an official notice stating that all animals brought into France had to be certified 'on pain of confiscation of the animal'. They also had to spend two months in quarantine when they arrived. Max realized they had to get out of there fast.

Max went to the cash desk with the bags of food. One of the girls working the checkout noticed Sidney. 'That's a beautiful Amazon blue. You're very lucky, young man. They are the best talkers in the world, apart from macaws.' She spoke quickly and at first Max had trouble lip-reading. 'You know you really shouldn't come into the shop with the parrot? I mean, suppose he was sick...I assume you have a licence for him?' She smiled. 'He's in very good health, you can tell.' Then she whispered: 'I know people who would pay a lot of money for a good talker. Does he talk?'

'Ask him!'

'Can you say something for me, eh? Can you?'

Sidney turned away. The woman behind the counter was annoyed. 'What's the matter with you?' She held out her hand. Sidney stared at her. 'He's got a nasty scar on his forehead. How long have you had him?'

'I have to go, my mother's waiting... and anyway, she said I wasn't to talk to strangers.'

The girl laughed. 'Very funny, very funny.'

Max hurried out. Back in the shop, the girl pulled a piece of paper from the pocket of her blouse and read the telephone number scribbled on it. She picked up the telephone and dialled. 'I can't, I'm at work. Yes, a young boy, he was about twelve with an Amazon blue. It was a beautiful bird.' A pause. 'Yes, blue feathers on its forehead and a scar.' Another pause. 'I don't know what colour hair he had.' A pause. 'Because he was wearing a hat, that's why. No, I can't keep him here, he's already left. What? OK, I'll see you when you get here.' She put the phone down and rushed out of the shop, looking up and down to see if she could see Max.

Max had crossed the street and was hiding behind one of the bookstalls. The bookseller was looking at him curiously. He watched as the angry girl from the shop went back inside. 'She asked a lot of questions', he whispered to Sidney. 'Let's get out of here. I bet she is trafficking in rare birds. That's it!' He had another thought. 'The two guys at the market, they were trafficking rare birds. She said herself that a good talker was worth a fortune. And you are certainly a good talker. You see, you are worth a fortune. The reason those guys were so angry was because they couldn't get you to talk. Suppose they'd already found a buyer for you and he'd already given them the money – they'd have to give it back now that I've got you. No wonder they were angry.'

As he walked back, Max thought about what he had learned in the pet shop. He might not know whether Sidney was male or female, or how old he was, but he knew that Sidney was an Amazon blue and an excellent talker. He also knew that he needed a certificate for him and that he didn't have one.

Barely a minute after Max and Sidney had left, a big Mercedes screeched to a halt outside the pet shop. There were two men inside. A big, stocky man climbed out and hurried inside.

CHAPTER 5

The Three Ages of Maths

A cast of thousands & a well-earned breakfast

Looking through the cataloguing system he had established, Mr Ruche set about making a list of things he needed to know. In short, he had to devise a list of the most important people in the history of mathematics. More than 2500 years of history in a single list. It was infuriating. He was desperate to unpack all the crates of books, but he realized that before he could organize the Rainforest Library properly, he would have to return to the National Library.

This time his work had to be planned like a military operation. Back in the National Library, he took out his notepad and opened it. It was already half-full. He was going to use a new fountain pen, a present from a customer in Venice. The pen, made entirely of glass, had been hand-blown in Murano.

He opened a bottle of ink and dipped the pen. Then he noticed that everything around him had gone quiet – everyone seemed to be looking at him strangely. It was only then that Mr Ruche noticed that no one else was using a pen. Everyone else had a laptop trailing cables. Luckily, he had a pile of huge mathematical dictionaries and histories of science with which he made a little wall to hide behind. He dipped the pen again and began to write. All around him nervous fingers tapped away at keyboards, the students itching to tell Mr Ruche how much better he could work if he had a computer. Mr Ruche ignored them. There was no time to lose: he needed to make some notes.

Section 1. Maths in Ancient Greece (700 BC to AD 700).

Sixth century BC, the founders: Thales (geometry); Pythagoras (arithmetic).

Fifth century BC, the followers of Pythagoras: Philolaus of Crotona, Hippasus of Metapontum, Hippocrates of Chios, Democritus (the atomist), the Eliatic School (from the town of Elia, in southern Italy), Parmenides and Zeno, Hippias of Elia (geometry).

Fourth century BC: the school of Athens. Plato and the work of the Academy.

Third century BC: the golden age of Greek mathematics. The three greats: Euclid, Apollonius of Alexandria, Archimedes of Syracuse. These three 'laid down the laws of geometry'.

After the third century BC (or thereabouts), almost all the important work in mathematics took place in Alexandria. This is called the Hellenic period. If Greek maths was born on Thales' journey to Egypt, it was to return there.

Third century BC: Eratosthenes (mathematician, astronomer, geographer, curator of the Library of Alexandria, he was the first to attempt to measure the circumference of the earth).

Second century BC: Hipparchus (whose work prefigures trigonometry); Theodosius (astronomer).

First century BC: Hero (mechanics).

Second century AD: Claudius Ptolemy (geographer and astronomer). Nicomachus, Theon of Smyrna (number theory), Menelaus (conic sections).

Third century AD: Diophantus (the beginnings of algebra).

Fourth century AD: Pappus (synthesis of geometric theory); Theon of Alexandria (geometry); his daughter Hypatia (the lone female mathematician of antiquity).

Fifth century AD: the 'great commentators' of Greek mathematics. Proclus (commented on the work of Euclid), Eutocius (commented on the work of Apollonius and Archimedes).

Sixth century AD: Boethius (last of antiquity's mathematicians).

It was getting dark. There were only two people left in the Library. All the seats in the great reading room were empty. Mr Ruche looked quickly at his notes and was surprised to find twenty names. A handful of men had been responsible for almost everything in Greek mathematics! His notes were brief, but sufficient for him to catalogue books of the period. Now all he had to do was compile two further lists covering the period from AD 500 to the present day. He decided he would stop in the year 1900. Only 1500 years to go!

On Tuesday morning Albert dropped Mr Ruche outside the National Library at nine o'clock. Mr Ruche had had the foresight to order the books he would need before he left the night before and so he could get straight to work on Section 2.

Section 2. Maths in the Arab world (from AD 800 to 1400)

Mr Ruche hesitated. This was virgin territory – he wasn't sure he could name a single Arab mathematician. He felt a sense of urgency as he plunged into the first book. He quickly discovered that the period referred to all mathematicians who had written in Arabic. There were Arabs and Persians, Jews and Berbers, scientists in the broadest sense of the word working not only in maths, but in medicine, astronomy, philosophy and physics. In this, they were very like the early Greek thinkers.

After a period of little activity spanning the fifth to the eight century, the knowledge developed by Greek thinkers began to filter into the Arab world. Once Arab mathematicians had absorbed it, they were to develop it further. From Alexandria via Byzantium, it came to Baghdad, the capital of the Muslim world.

Arab scientists, particularly in the ninth and tenth centuries, were accomplished translators as well as mathematicians. It was they who began the enormous task of translating Greek mathematics: Euclid, Archimedes, Apollonius, Menelaus...In doing so, they absorbed the wisdom of the ancient Greeks. But they also opened up the field of mathematics, developing new disciplines of which the Greeks were unaware. And they assimilated ideas and works from other countries, particularly from India.

Like the early Greeks, Arab mathematicians were interested in medicine, astronomy, philosophy and physics, and this interest in other disciplines would inform the development of algebra and trigonometry.

Ninth century AD

Baghdad: Al-Khwārizmī (algebra, first- and second-degree equations).
Egypt: Abū Khāmil (broadened the scope of algebra to include multiple

equations with multiple variables); al-Karajī (the first mathematician to consider irrational numbers). Al-Fārisī laid down the basics of elementary theory of numbers and established that 'every number is composed of a finite number of prime factors, of which it is the product.'

Late ninth century

Baghdad: Three brothers called Banū-Mūsā (geometry); Thābit ibn Qurra, Al-Nayrīzī and Abū-l-Wafa (calculated areas: parabola, ellipse, the theory of fractions, devised the sine table and developed trigonometry as an independent discipline).

Late tenth century

Al-Bīrūnī (geographer, astronomer and physicist), Ibn al-Haytham (number theory, geometry, the concept of infinity, optics, astronomy – but no algebra).

Ibn al-Khawwām: first to present the equation which is to become Fermat's famous Last Theorem: a cube cannot be the sum of two cubes: i.e. $x^3 + y^3 = z^3$ cannot have a solution in whole numbers. Two other great mathematicians: al-Karajī and al-Samaw'al, who in the twelfth century developed his work. Al-Samaw'al proposes a system of 210 equations with ten unknowns. The result: the arithmetization of algebra.

Mr Ruche felt he needed to expand on this.

The arithmetization of algebra involves applying arithmetic operations (addition, subtraction, multiplication, division, square root) to variables. Previously these had been applied only to numbers. In applying them to variables, calculation expanded beyond simple numbers and into algebraic terms.

Al-Karajī studied the terms x^n and $1/x^n$. Al-Samaw'al considered negative exponents, and demonstrated the fundamental rule that $x^m x^n = x^{(m+n)}$. He was the first person to use *recurrence* to establish mathematical results, principally in number theory. Calculating the sum of n prime numbers from the sum of their cubes.

In the margin of his notebook, Mr Ruche started jotting down 1 + 2 + ..., but he didn't have enough space. He went back to the page itself and wrote:

$$1+2+3+\cdots+n=\frac{n(n+1)}{2}$$

He felt an urge to test the formula and tried it for $n = 5$. Adding the numbers 1 to 5 gave him 15; using the equation, it produced:

$$\frac{5\times(5+1)}{2}=\frac{5\times(6)}{2}=\frac{30}{2}=15$$

The equation worked!

He moved on to the next formula, which seemed a lot more difficult: the sum of the squares of n whole numbers is

$$1+2^2+3^2+\cdots+n^2=\frac{n(n+1)(2n+1)}{6}$$

The sum of the cubes of n whole numbers is equal to the sum of those numbers squared:

$$1+2^3+3^3+\cdots+n^3=(1+2+3+\cdots+n)^2$$

'I'm wasting time here', thought Mr Ruche. 'I can't check every formula in the history of maths!' He decided to take a break. He needed a coffee, but not one from a machine. A real heavy-duty espresso. He wheeled himself to the bar at the end of the road, and returned reinvigorated. Back at his desk, he looked around for his fountain pen and couldn't see it anywhere. The computer chain-gang looked at him coldly. He hunted around. It certainly wasn't on his desk. As he leaned over to look under the desk, he spotted a bulge in one of the maths books – his pen had slipped between two pages. He picked it up and ran his finger lovingly along the twists and turns of the barrel, and then began to write:

Late eleventh century: Omar al-Khayyām (mathematician and poet, a great algebra man).

Late twelfth century: Sharaf al-Dīn al-Tūsī (another great algebra man. He was using something very like the idea of derivatives five hundred years before anyone in the West).

Thirteenth century: Nasīr al-Dīn al-Tūsī (astronomer, who revised the Ptolemaic world-view).

As he wrote this last name he realized he had seen it somewhere before, but he was in too much of a hurry to stop and check.

Early fourteenth century: al-Kāshī (director of the observatory of Samarkand, wrote a synthesis of mathematics in the Arab world charting the links between algebra and geometry, algebra and number theory).

It was 7.45 p.m. by the time he finished Section 2. He had done all he could. Hopefully, his notes would be sufficient to catalogue the books on Arab mathematics in the Rainforest Library. Tomorrow he would tackle Section 3 – maths in the Western world from the fifteenth century to 1900. Mr Ruche wheeled himself out of the reading room. Outside, the air was chill and damp. He had to wait for a while before he caught a taxi.

Mr Ruche hadn't understood everything he was jotting down in his notepad. If he didn't have the faintest idea about something, he simply copied word for word. This chronology of maths would not magically help him understand all the ideas – it wasn't intended to. He simply wanted to familiarize himself with the subjects, so that later he could deal intelligently with the major themes of mathematics.

His notes posed simple questions: What were the great ideas of a particular period? What subjects were important? Who were the important mathematicians? Which problems posed by earlier thinkers were now resolved? What new questions had been thrown up by new work? Were there new disciplines? Mr Ruche needed only as much as would satisfy an enlightened amateur.

But – and this was the big question – in a subject like maths, could there be such a thing as an enlightened amateur? He did not try to answer that. He had a much simpler aim: to sort out the crates and catalogue Grosrouvre's books for the Rainforest Library.

The following morning, Mr Ruche's whole body ached and he was running a temperature. He had caught a nasty dose of the flu, probably while he was waiting for a taxi outside the National Library. Perrette phoned Albert to let him know that Mr Ruche would be staying at home. She had only seen Mr Ruche ill twice in all the time she had known him.

On the third day he felt a little better. Still coughing and sniffling, he wrapped himself up warmly and Albert drove him down to the library, where he went straight to the reading room. He took out his pens and notebooks. He wrote:

Section 3. Mathematics in the Western world (1500–1900)

The category seemed enormous – he would probably have to subdivide it later. For the moment, it would do. He would approach the task geographically: Italy, then France, Britain and Germany, the Netherlands, Switzerland, Russia, Hungary and Poland. He began:

Sixteenth century

The golden age of elementary algebra. The Italian school in Bologna (third- and fourth-degree equations): Tartaglia, Cardano, Ferrari, Bombelli. (Discovery of complex numbers.) Great progress in symbolic notation: Viète, Stevin.

Seventeenth century

Invention of logarithms: Napier (Baroque mathematics).

Algebra: Girard, Harriot, Oughtred.

Analytic geometry (which established the link between space and numbers using algebra): Fermat, Descartes, Geometry of indivisible: Cavalieri, Roberval, Fermat, Gregory de Saint-Vincent.

Calculus – differentiation and integration: Newton, Leibniz, Jacques and Jean Bernoulli, Taylor, Maclaurin.

Number theory: Fermat.

Probability: Pascal, Fermat, Jacques Bernoulli.

Geometry: Desargues, Pascal, La Hire. . .

His head was spinning, He was too old for all this. He felt like going home and having a lie-down. He closed his eyes and remembered a time, over sixty years ago, when he could spend every waking hour cramming for his exams, but he wasn't twenty years old any more – he was old and he was ill. But the thought of all the books crammed into crates gave him a burst of energy.

Eighteenth century

Classicism. The golden age of analysis; after numbers and shapes, *functions* became the new rage in mathematics.

Differential equations, the study of curves, complex numbers, theory of equations, trigonometry of spheres, calculation of probability, mechanics: the Bernoulli brothers, Euler, D'Alembert, Clairaut, Moivre, Cramer, Monge, Lagrange, Laplace, Legendre.

Problems posed earlier by Leibniz and Newton are resolved; squaring the circle and calculus have made great progress.

Only one century to go!

Nineteenth century

The dawn of a number of new disciplines in mathematics (groups, matrices...).

The theory of functions using imaginary numbers: Cauchy, Riemann, Weierstrass.

Algebra: Abel, Galois, Jacobi, Kummer.

Geometry: Poncelet, Chasles, Klein, Gauss.

Non-Euclidean geometry: Gauss, Lobachevsky, Bolyai, Riemann.

Matrices: Cayley.

Algebra: Boole.

Set theory: Cantor, Dedekind. And Hilbert and...

Enough! He was certain he had missed others, but his head felt as though it was about to explode. Mr Ruche had used up three packets of tissues and written ten pages. He was exhausted, but he had 2500 years of mathematical history in his hands.

Perrette and Mr Ruche set aside the weekend to catalogue the Rainforest Library. Mr Ruche rolled his wheelchair to the first crate, lifted the lid, took out a book and solemnly announced: '*Introductio in analysin infinitorum*, by Euler, for Section 3!' The first book of the Rainforest Library was placed on its shelf. The next volume was *Arithmetica* by Diophantus, which went into Section 1. The first crate was soon emptied and put out into the courtyard. The shelves began to fill up.

They were both surprised by the number of recently published books and had to revise their whole idea of the Library – it was not only a collector's library, but a research library. There were magazines too – thousands of them. Perrette sealed up the crate full of magazines and pushed it against the far wall. They went back to the books.

The Arithmetic of Elliptical Curves, Silverman, Section 3; *In artem analyticem isagoge*, Viète, Section 3; *Traité sur le quadrilatère complet*, Nasīr al-Dīn al-Tūsī, Section 2: *Mirifici logarithmorum*, Napier, Section 3; *Disquisitiones arithmeticae*, Gauss, Section 3; *Miftāh al-hisāb*, 'The Key to Arithmetic', al-Kāshī, Section 2; *Spherics*, Menelaus, Section 1.

Monday morning came, and they still hadn't finished. Before opening the bookshop Perrette went to the studio, where she found Mr Ruche, surrounded by boxes, asleep in his wheelchair, the tartan rug fallen from his knees. He looked happy. His breath rose and fell like wind in a sail. She thought for a moment how the accident had aged him. She left him sleeping.

Mr Ruche studied the book in his hands. He had never heard of the author, and he couldn't make head or tail of the contents page. He leafed through the book again. A piece of paper slipped out and floated down under a bookshelf. He couldn't reach it himself but was embarrassed to ask for help. He thought for a bit and then smiled, wheeled himself over to the cupboard, took out the vacuum cleaner and plugged it in. After a little rummaging around, he heard the sound of something being sucked onto the nozzle. It was a piece of card. Mr Ruche recognized Grosrouvre's handwriting on it – thin and spidery and always in black ink, just like the letter he had sent. It was a synopsis of the book, with a commentary by Grosrouvre. He pulled out some more books. At the back of each of them, held on with an elastic band, was a similar card. The cards would be a great help in sorting out the books.

After work, Perrette ate quickly before joining Mr Ruche in the library. There were more empty crates now than full ones. Soon, there was only one full crate left. Like all of those that went before them, the books in the last crate took their place on the shelves of the Rainforest Library.

By the time Perrette took the last crate out into the yard, dawn was breaking. They had never seen so many old and rare books in one place before. Mr Ruche was proud of his old friend. Grosrouvre was probably the only man in the world who could have amassed such a library. Nearly every book was a first edition, some of them four or five hundred years old. There were even some *incunabula* –

the earliest printed books – dating from before 1500. Some of the books had notes and annotations, others had beautiful hand-drawn plates. Each and every one was in impeccable condition. The bindings were original, though here and there a spine had been broken by book lovers poring over them. On the RFL's shelves were thousands of books, in Greek, Latin, Arabic, Italian, German, English, Russian, Spanish and French. It was a mathematical tower of Babel.

'In the crates I've sent, I think you'll find the world's best collection of books about mathematics. Everything that should be there is there. It's certainly the most complete private collection on the subject ever made.'

Grosrouvre was right. He might embroider a story now and again, but if he told you something unbelievable, you could be certain it would turn out to be true. They closed the door to the Rainforest Library and went to the brasserie on the corner for breakfast.

CHAPTER 6

Friends and Enemies

A second letter; a suspicious death & Fermat's Last Theorem

Perrette went into Mr Ruche's room and handed the letter to him through the curtains of his four-poster bed. 'It's from Grosrouvre!' he shouted. 'Didn't I tell you, Perrette? Didn't I say he'd write again soon?' He threw back the curtains of his bed. Looking closely at the letter, Mr Ruche noticed it had a letter-head: Manaus Police Department: Amazon. It wasn't from Grosrouvre after all. Disappointed, he opened the envelope as Perrette opened the windows onto the courtyard.

'Shit, shit, shit!'

Perrette spun round, shocked – she had never heard Mr Ruche swear. He looked at her and his face fell. He handed the letter to her. It was from the Police Commissioner in Manaus. In faltering English, Commissioner Grindérios regretted to inform Mr Ruche that Señor Elgar Grosrouvre had died in a fire at his home. His charred body had been recovered in the remains. A letter to Mr Ruche had survived the fire, and the Commissioner had forwarded it.

Mr Ruche took out the slightly scorched envelope; the writing was certainly Grosrouvre's. He propped himself up on his pillows and Perrette came and sat on the edge of the bed. A look of pain flickered across Mr Ruche's face as he opened the envelope. 'Just like him to go and die just when we'd got in touch again.' He couldn't bring himself to read the letter, and Perrette gently took it from him and began to read:

Manaus, September 1992

Dear πR,

I have only a couple of hours left. Just enough time to write and explain. I owe you that, at least. First you'll want to know why I came to the Amazon. The truth is, I felt suffocated in Europe. You should know how much I have always needed fresh air. Where better to come to than 'the lungs of the world' – the greatest oxygen reserve on the planet – the rainforest?

When I left Paris the one thing I had in mind was a sixteenth-century Portuguese proverb: 'south of the equator, there is no such thing as sin'. Well, if you look at a map, you'll see that Manaus is just south of the equator – no more than two or three degrees. When I moved, I moved house and country and hemisphere.

When I arrived in Manaus I worked for a while as a labourer. I spent whole days in the forest and never met another soul. It was then that I had the idea – an idea that never left me and one that helped me survive through the bad times. I decided to set out to prove one of the most famous mathematical theorems in the world. I don't suppose that will mean much to you, but it was a huge undertaking.

Why did I decide to do it? Was I trying to compete with the great thinkers of mathematics? No, I never was very competitive. Did I want fame? Of course not. No, πR, I did it to survive. You can't imagine the sheer force of nature out here. It's frightening sometimes. How can a man stand up to a force which devours everything and which nothing can destroy? Here life is abundant, overpowering. It is paradoxical: everything is slowly rotting and yet the atmosphere is so full of life it makes death seem that much closer. In this intensely physical world, to escape from this cycle of growth and death I steeped myself in pure, unchanging ideas.

You will never see a mathematical definition rot, or a theorem die. I chose maths, not only because I had studied it at college but because it is permanent. Faced with the frightening force of nature, I found peace in the purest, simplest ideas. I asked myself what I should study. In the end I chose a handful of mathematical theorems. Fermat's famous Last Theorem, Goldbach's conjecture, and others.

Imagine knowing that a continent exists, but never being able to reach it. An unproved theorem – a conjecture – is a simple statement, something the whole world takes for granted, but which no one has been able to prove. That was the sort of challenge I needed.

I chose to work on two specific theorems. I spent all my time working, day and night. More night than day. In the end, I succeeded in proving both. The oldest and the most famous, Fermat's Last Theorem and Goldbach's conjecture. And the solutions? When you write them down they are so simple that even you could understand them, πR.

If word got out about this, it would be on the front page of every newspaper, but I've decided not to publish, to keep the proofs secret. I don't intend to tell anyone about my work.

Are you shocked by my silence? Though I don't have much time left, I'll try to explain. Although we're very different, I know you'll understand. You should know that I would not be the first person in the history of maths to be secretive, though secrecy is no longer much in fashion. Nowadays people rush into print before they've even finished the proof.

As you will remember well, we never agreed on anything – I've always thought that was what kept our friendship alive. I loved Aristotle, who left behind a library of books; you loved Socrates, though not a single word of his survives. I loved Danton, because he knew how to compromise; you liked Robespierre because he never allowed himself to be tainted. You loved Rimbaud, though you've never left Paris in your life; I loved Verlaine and I'm the one who went to the ends of the earth. But together we loved so many things.

You always said that philosophy has two origins. You always preferred Thales, while I preferred Pythagoras. Both travelled to Egypt and down the banks of the Nile. Thales came back with a tale of shadows and Pythagoras with a story of numbers, which I talked about with you so often.

Pythagoras talked to animals. Did you know that he once convinced a bear to stop terrorizing a local village? I've adopted a lot of animals since I've been here and, of course, we talk all the time. We've had many fascinating discussions.

You know that Pythagoras founded a sort of sect, I suppose you would call it. One of its rules was never to divulge the things its members had learned. To stop their secrets from falling into the wrong hands, Pythagoreans rarely wrote anything down and their knowledge was passed on by word of mouth. What is written down is permanent, while the spoken word is ephemeral. To make sure their wisdom did not disappear, they developed exercises to train the memory.

One of the akousmata who belonged to the sect, Hippasus of Metapontum – an excellent mathematician – revealed their discovery of irrational numbers and their incredible properties. It was a discovery to which he had contributed. He died shortly after under suspicious circumstances.

There are people – people I have done business with – who have found out about my work. People who are not what you might call pacifists. They offered me a lot of money for my proofs and I refused. It would be hard to keep something from these men if they truly wanted it. They'll be back tonight.

Believe me, πR, I won't give them the proofs. I will burn them as soon as I have finished this letter. In case something should happen to me, I would not want them to be lost forever, so – like the Pythagoreans – I have confided them to a loyal friend who, I know, will remember them.

I do not wish to keep secrets from you. If I remember, when we were young, whenever I kept secrets from you, you always managed to work them out. I think I've said enough.

Remember your old friend Thales? He was a shopkeeper before he became interested in mathematics. I'm sure your bookshop is doing well. You have always been very good at selling something you believed in.

You should have received my books by now – it is a magnificent library, isn't it? I've only just realized that I forgot to send you the note of how I had classified them, but I'm sure you don't need it now. I'm sure you've already classified them in your own way.

It's almost dark, I had better get ready.

Love

Elgar

P.S. Do you know what first interested me about Pythagoras? He invented the word 'friendship'. When someone asked him what a friend was, he replied 'Someone who is another me, like the numbers 220 and 284.' Two numbers are 'friends' if each is the sum of everything that measures the other. The most famous of these so-called 'amicable numbers' are 220 and 284. They're a fine couple. You should try checking them sometime.

What about us – are we 'friends'? What is the sum of those things that define you or me? I think perhaps the time is coming when we will find out.

Perrette put the letter on the bedside table. Mr Ruche lay staring at the canopy of the bed. She left the room without saying a word. He didn't hear the door close behind her.

'That was Grosrouvre all over', thought Mr Ruche. 'No news for fifty years, and when he finally gets in touch to let me know he's still alive it's only to let me know that he won't be soon. I did my mourning for him years ago and now he comes along and opens up a wound I thought was closed for ever.'

Mr Ruche took a little longer than usual to get dressed. From his wardrobe he chose a pair of black moccasins, the kind you might wear to a funeral, and polished them carefully. Grosrouvre had been his only true friend and now he had lost him for a second time. This time, it was for good.

He bent over to tie his shoelaces, then suddenly sat bolt upright, furious. If Grosrouvre hadn't sent him the books, they would all have been burned in the fire. This shook him – all the books, destroyed. These books, which he had spent days in the library cataloguing and shelving, they were priceless. Mr Ruche smiled – Grosrouvre's library had escaped destruction twice in as many weeks. First – if the delivery man was to be believed – from sinking to the bottom of the Atlantic and then from the flames of a fire in the Amazon. It was a miracle. Unless – a thought suddenly occurred to Mr Ruche – there was a connection between Grosrouvre sending him the books and the fire. Perhaps Grosrouvre had sent him the books so that they *wouldn't* be destroyed. But that would mean that he knew about the fire, that he knew weeks in advance that his house was going to burn down. Did he foresee it, or was it planned and, if so, by whom? Mr Ruche could hardly bring himself to entertain the idea. It was easier to hope that it was simply a coincidence, that it was a stroke of luck that the books had been sent in time to escape the inferno.

Mr Ruche went out for his constitutional. He passed the church, crossed the Place des Abbesses and wheeled himself onto the terrace of a café. It was a fine afternoon. Mothers pushing buggies along, the usual trio of tramps sitting on the bench, a couple of tourists talking excitedly about the art nouveau entrance to the métro. Some of the regulars nodded to him and he nodded back. Something in his face told them that he didn't want company.

He was surprised to hear himself order a brandy and water. When the waiter arrived with the brandy glass, he remembered why. He and Grosrouvre always drank it on special occasions. It had been their favourite drink. Today, Mr Ruche was drinking it in mourning. He sipped it slowly, a hundred questions humming inside his head – about Grosrouvre's death, but also about the references to maths in his last letter. He was certain that his friend had included them for a reason. He would have to look into it closely, and study Pythagoras with the same diligence as he had studied Thales.

The square was quiet and the afternoon warm and sunny. There were few people and few cars. It was an ideal place to reminisce. Grosrouvre had been right – they had never agreed on anything. It was as if they had decided to divide the world into two: you take this; I'll take that. Grosrouvre used to say, 'I am me and you are you and we are not them.' 'Grosrouvre always talked in formulas', Mr Ruche remembered. 'It didn't really endear us to the other students, but we didn't care.'

Mr Ruche had always been struck by the sheer physical strength of Grosrouvre. He had noticed it in the army. It was 1939 and they had just been called up. 'Built like a Norman castle. That's what I used to say', thought Mr Ruche. 'When Grosrouvre went dancing, the girls always had to rest their head on his chest. Head and shoulders above them, Grosrouvre moved through the other dancers like the prow of a ship.' Mr Ruche stopped. He didn't want to remember any more.

He asked the waiter to bring him a pen and paper. He leaned forward and began to write. Angrily, he scribbled things down and crossed them out. After a lot of false starts, he ended up with this:

Factors of 220: 1, 2, 4, 5, 10,11, 22, 44, 55, 110
Factors of 284: 1, 2, 4, 71, 142

He added up the factors of 220, crossed it out and began again. The result: 284. Mr Ruche smiled – he was halfway there. He totted up the factors of 284: 220. 'Well, there you go...they're definitely "friends".'

Perrette arrived and sat down at the table with Mr Ruche. She noticed the brandy glass, and ordered a sherry.

'We've never really talked, Mr Ruche.'

He looked at her for a long time. She had barely changed since the day, seventeen years ago, when she had walked into the bookshop. Her curly hair was shorter, but black as charcoal still. No one would have thought she was forty. 'That's true', he admitted. Then, after a moment, 'Would you mind calling me Pierre?'

'Oh, I couldn't do that!' she said, blushing. 'I don't think you like familiarity very much. It's precisely the formality of our relationship that makes us so close.'

'No one has ever said that to me before, but you're right.'

'A lot of things have happened since I came to live in the rue Ravignan. I think we've come to a turning point in our...' – she searched for the right word – '...in the life we share together. I think we have to be careful.'

Mr Ruche listened. He had never heard her talk like this before.

She went on, 'This thing with Grosrouvre is complicated and I don't think you can deal with it on your own. I know you're not asking for help, you never do. I know Grosrouvre was your friend, and I would like to have met him, but do you know who he reminds me of? The American Uncle, like in the film. You know what I mean – the one who goes off to see the world as a young man, lives a wild life. Nobody ever hears of him again, until one day a lawyer arrives with a will and says that he has left you a fortune. Only in your case, you got the fortune before he died: his library.' Her eyes were shining now. 'It's more than a fortune, it's priceless. And that letter this morning, it was a will, in all but name.'

Mr Ruche looked up. She was looking at him mischievously. He shrugged his shoulders. 'Though I think it's something of a poisoned chalice,' added Perrette, 'but the kids will help us to sort it out. They're a clever bunch, you know. And I'm not exactly dim myself.'

Perrette put her hand on his, and they agreed to convene a general meeting that evening. Perrette knew as little about Mr Ruche as he did about her. They were both very private people. Only in the last few days had they seen glimpses of each other's lives.

Suddenly she asked him, 'Why were you so close to Grosrouvre?'

Mr Ruche's face changed. Her question seemed to have transported him to a different time, a different place. 'The Germans attacked...we were taken by surprise. Grosrouvre managed to escape, I was caught. Days later, I saw him being brought into the camp. He was limping badly. His leg had been broken in an ambush. It was winter and very, very cold. I caught pneumonia. There wasn't any medicine. But Grosrouvre managed to get hold of some mustard from God knows where, and made a mustard poultice. It burned horribly, and I was constantly shivering. He gave me his coat and watched over me day and night. I was delirious. When I regained consciousness, he was sitting on a stool beside my bed. He said, "Philosophy goes on for ever, so hurry up and get better – people are counting on you." I didn't know who he meant, but then he listed all the philosophers he knew I admired.

'I spent a long time convalescing...I was as thin as a rake and he used to say, "Well, as we're right out of mustard, we're dead if we catch anything else. So, since I'm up and about again, I propose we leave this marvellous party." And we did – we escaped. We split up so as not to get caught. I went across the fields and he went through the woods. That was the last time I ever saw him.'

That evening, after dinner, Mr Ruche tapped the table with his glass to bring the meeting to order. Max was sitting opposite his mother so that he could lip-read. Stuffed full of sunflower seeds and honey, Sidney was dozing on his perch. Jon-and-Lea were on the sofa.

With her back to the fireplace, Perrette stood and read the letter. She read slowly, stopping from time to time so that they could all weigh Grosrouvre's words carefully. When she came to the last sentence – 'What is the sum of those things that define you or me? I think perhaps the time is coming when we will find out' – everyone started talking at once, about the fire, about Pythagoras, about theorems and proofs and how strange Grosrouvre seemed. Perrette handed the letter to Mr Ruche. In the middle of all the clamour, they heard Max say, 'Those guys are pretty nasty.'

From his lips, it sounded like the worst abomination. Max turned to Mr Ruche:

'If your friend didn't want to sell his...his...'

'His proofs', offered Perrette.

'...then he didn't have to. They belonged to him, he was the one who did them. Nobody could force him to. I think they're to blame for the accident.'

'What makes you think it was an accident?' asked Jonathan.

'It was an accident,' interrupted Mr Ruche. 'I've been thinking about it since this morning. In a way, I feel I'm to blame.'

'What are you talking about?' asked Perrette angrily. 'How could you be responsible for an accident on the other side of the world?'

'It's not about distance, Perrette. Once he'd decided to burn his proofs, he sat down and wrote this letter. Eight pages of it. He didn't notice the time passing. He says himself that by the time he had finished it was dark; he only had a few minutes left. They were coming. Maybe he was careless, maybe that's how the house caught fire. I can see him there, watching years of work go up in smoke. It must have been terrible. He couldn't run – he's...well, he's not a young man any more. I don't know...maybe he had a turn, knocked something over...'

'I don't think it was anything like that', Jonathan said gently. 'I don't think you're to blame for anything.' Mr Ruche nodded sadly. 'I think your friend arranged everything', Jonathan went on. 'I think the letter he sent you is his will. I think he planned his own death.'

'You mean...'

'That he committed suicide' said Jonathan.

'Grosrouvre would never do that', protested Mr Ruche.

'Mr Ruche, listen. Grosrouvre had already decided to refuse their offer. He had destroyed what the guys had come for. He knew them, he knew what they were capable of. Let's say they show up at his house, and he says, "I've just burned everything you came for." How do you think they'd react? They'd be furious – they'd be sure to try to get the proofs out of him – after all, he would still remember them. They might well beat him up, torture him. No, I think Grosrouvre knew exactly what was going to happen, so he wrote the letter, burned his papers and set fire to the house. He killed himself. I don't know how, there must be a way. Doesn't curare poison come from South America?'

'But why would he kill himself?' asked Perrette, 'why not just take off?'

'Because he knew those guys. He knew that wherever he went, they would catch up with him. They must have been an organized gang.'

'Are you working on a plot for a movie or something?' Lea

interrupted suddenly. 'It doesn't matter who they were, what difference does it make how it happened?'

Jonathan ignored his sister and stood up, shaking out his long hair. 'He sent you the books because he knew he was going to set fire to the house. He would never have been able to burn them. He could burn his proofs, because they were his to burn: he wrote them, but not the books. You don't send a whole library halfway round the world for no reason.'

Lea got up and, without a word, went upstairs. She sat on her bed in the attic fuming. Here they were spending a whole evening trying to work out how some old man in Manaus died and nobody seemed to care how she and Jonathan were born. Even Jonathan didn't seem to care. Why was it more important to find out how someone died in some hole in the back of beyond than to find out how they were born in a hole right here in Paris?

Downstairs, Max had another theory. 'Maybe he sent you the books to make sure these guys didn't get them. Maybe he thought they'd try to threaten him: if he didn't give them the proofs, they'd burn the books.'

'Perhaps,' said Mr Ruche, 'sending me the books proves nothing at all.'

'If someone is dead, there are only four possibilities', said Perrette confidently. 'Natural causes, accidental death, suicide or murder. We know it wasn't natural and you've talked about accidents and suicide, but what if he was murdered?'

They looked at one another, shocked. Not one of them had thought about murder. Everyone was silent. Mr Ruche sat up in his wheelchair.

'They had no reason to kill him', said Jonathan. 'They needed him alive, especially if he'd burned the proofs. He was no use to them dead.'

Mr Ruche listened. The casual way they were talking about his friend's death was hurting him deeply.

'True,' agreed Perrette. 'Maybe Jonathan was right – maybe they were trying to get him to talk. Grosrouvre refused. They threatened him, but he stood up to them and they killed him.'

Perrette's story was plausible, but Jonathan couldn't help pressing her: 'But, in that case, why burn the house down?'

'To cover up the fact that they'd killed him, or to cover their tracks.'

It was late. Everyone was silent, each weighing up what had been said. Mr Ruche was sure it was an accident; Jonathan thought it was suicide; Perrette, murder; and Lea clearly didn't care. Max didn't have an opinion about how Mr Ruche's friend had died, but he was sure of one thing: those guys were to blame. The important thing, everyone agreed, was to find out why they were so interested in Grosrouvre's proofs. What use, they wondered, could a couple of unpublished maths theorems be to them? And who was the loyal friend to whom Grosrouvre had explained his proofs? Whoever it was would have to have an extraordinary memory.

CHAPTER 7

The Numbers Game

The school of Pythagoras; a cult of secrecy & the music of numbers

Mr Ruche was convinced that there was more to Grosrouvre's letter than met the eye. He knew that if only he could read between the lines, he could decode it. Everything seemed to revolve around Pythagoras. Why had Grosrouvre mentioned him, and what was he trying to tell Mr Ruche?

Mr Ruche realized he would have to make a study of the life and work of Pythagoras and the Greek thinkers who were part of his school. What was the *akousmata* he mentioned in the letter, and why were they all sworn to secrecy? Why was the discovery of irrational numbers so 'incredible' and how could it be important enough to cause the death of Hippasus of Metapontum? Did it have anything to do with Pythagoras' famous theorem? Mr Ruche had flirted with some of these ideas as a young man, but had only a vague recollection of them now. Grosrouvre had been right when he said that Mr Ruche had never really been interested in Pythagoras and his school. He had found them too mystical, too religious for his taste.

Mr Ruche went into the Rainforest Library and wheeled himself along the shelves to the section on Greek maths. He took the long pincers he used for reaching books and brought down *The Life of Pythagoras* by Iamblicus, written in the second century AD. He wheeled himself over to the desk he had installed in a corner of the studio – a beautiful old leather-topped escritoire with carved legs and balled feet.

He read *The Life of Pythagoras* at one sitting. It was as gripping as a novel. The cover was worn; it must have been consulted often by Grosrouvre and its previous owners. Some of the pages were dog-eared, and these he read with particular attention. He took out his notebook and his glass fountain pen and wrote:

Pythagoras invented the word 'philosophy'.

As in the case of Thales, read Mr Ruche, no written work of Pythagoras survives. Even the dates of his birth and death are unknown. What is known, however, is that he was born on the island of Samos in the Aegean sea in the 6th century BC, and died in the town of Crotona in southern Italy. He was only eighteen when he participated in the Olympic Games, where he won every single boxing tournament. After that, he decided to travel. He went first to nearby Ionia, where he spent some years with Thales and his pupil Anaximander. Then he travelled to Syria, where he stayed with the Phoenician sages. From there he went to Mount Carmel in what we now call the Lebanon, and then to Egypt, where he stayed for twenty years. In the temples on the banks of the Nile he learned the wisdom of the Egyptian high priests.

When the Persians invaded Egypt, he was taken prisoner and brought to Babylon. He spent twelve years there in the capital of Mesopotamia, where he learned much from the scribes and from the wise men. He returned to Samos forty years after he had left, older and wiser.

At that time, Samos was ruled by the tyrant Polycrates and, unable to bear his tyranny, Pythagoras left once again. This time he travelled west, towards the coast of Greece. From there he went to Sybaris in southern Italy – famous throughout the ancient world as a city of pleasure – but it was nearby in the town of Crotona, that Pythagoras made his home and founded his 'school', which lasted for one hundred and fifty years. In its time, it nurtured 218 Pythagoreans, including Hippocrates of Chios, Theodorus of Cyrene, Philolaus, Archytas of Tarentum and, of course, Hippasus.

Mr Ruche put aside the biography and opened the other books dealing with Pythagoras' mathematical work and that of his disciples. Hippasus was among the first of them; he was the leader of the *akousmata* – the 'acousmaticians' were candidates to be initiated into the school, while Pythagoras was leader of the 'mathematicians' – those who had been initiated.

Hippasus is associated with the discovery of the third type of mean. Means are numbers which indicate the average of (originally) two numbers. To the arithmetic mean and the geometric mean was added a third, called the harmonic mean. It would become the basis of the study of harmonics in music. Pythagoras was right: numbers truly were everywhere.

The arithmetic mean is usually just called the mean. It is half the sum of two numbers: 'The difference between the first number and the second is exactly equal to the difference between the second and the third.' Mr Ruche wrote down the formula:

$$a - b = b - c$$

b is the arithmetic mean of *a* and *c*:

$$b = \frac{1}{2}(a + c)$$

From Hippasus, Mr Ruche moved on to another of Pythagoras' pupils, Hippocrates. Hippocrates studied crescents, known in maths as *lunes*. It was Hippocrates who first established the quadrature of lunes – quadrature, or 'squaring', involved calculating a square which would be equal to the surface area of the crescent. Hippocrates' lune was the first quadratic figure of a curved object. Mr Ruche began a new page and drew the following:

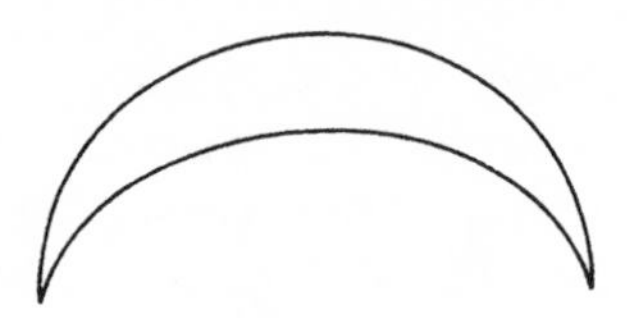

He made a note in the margin:

> The three great problems of Greek mathematics: squaring the circle (quadrature), duplication of the cube and the trisection of an angle.

As a young man, Hippocrates had squandered all his money. As an old man, he was thrown out of the Pythagorean school for 'demonstrating geometry for money'. It was precisely what Grosrouvre had refused to do – accept money for his mathematical proofs. If he had, he would still be alive today. Grosrouvre wasn't prepared to give his work away, as Hippasus had, or sell it, as Hippocrates had.

Mr Ruche continued reading: 'The school was founded in the town of Crotona in the tip of southern Italy. A rich, powerful man named Cylon who lived in the town wanted to be admitted to the Pythagorean school, but was rejected. Cylon was an autocratic man: he was not in the habit of being refused something he wanted.'

Mr Ruche stopped. The last sentence seemed familiar. Then, he remembered Grosrouvre's letter: *'It would be hard to keep something from these men if they have decided they want it.'*

He went back to his book. 'Furious at his refusal, Cylon decided to take his revenge. The members of the Pythagorean school met regularly to discuss city politics. One night, Cylon and his supporters set fire to the meeting house. All but one of those inside died in the blaze.'

Mr Ruche shuddered. This was no coincidence. Grosrouvre's 'business partners', like Cylon, had burned what they could not possess. Mr Ruche was angry now. Perhaps Perrette was right, and the fire in Manaus had been set deliberately. If so, he had to find the 'Cylon' who had murdered Grosrouvre. To do that he had to concentrate on the maths Grosrouvre mentioned in his letters – he was convinced all of the answers were in there.

He picked up the biography again. 'One man survived the fire in Crotona: Philolaus. An astronomer and cosmographer, Philolaus had proposed an astounding theory about the world two thousand years before Copernicus and Galileo. Not only did the earth turn, according to Philolaus, but it was not the centre of the universe. Philolaus further suggested that there was a fire at the centre of the universe around which the earth, the planets and even the sun turned.' Mr Ruche wondered whether the theory came before or after the fire from which he had miraculously escaped.

He was astonished by the next sentence: 'Across a small strait from Crotona in southern Italy was the town of Tarentum, where Archytas invented the number 1.' Mr Ruche paused for a moment. Surely the number 1 had always existed? According to this book, it seemed that Greek thinkers believed numbers began with 2. To them, there was 'one' and 'more than one'. They believed that 'one' was not a quantity but a statement of existence, whereas numbers were about multiplicity: 'one is that which is'. Mr Ruche was excited – this was pure philosophy. In taking the singularity of 'one' and its epistemology, Archytas made a number of it.

Mr Ruche continued to take copious notes. Archytas was not only the 'father of one' but the 'first engineer'. Marshalling his knowledge of mathematics and geometry, he developed a theory of mechanics and is reputed to have made a mechanical bird – a wooden dove which could fly, powered by a small engine.

There was something else: Archytas was the first 'graffiti artist' in history. He wrote graffiti because he could not bring himself to swear. When he felt he absolutely had to, he simply wrote the offending word on a nearby wall. He reminded Mr Ruche of Max, who never swore. It was as if words were too important for him to waste them.

For Mr Ruche, however, Archytas' crowning glory was that he had saved Plato from Denys, the tyrannical ruler of Syracuse who wanted the philosopher assassinated. Archytas dispatched a battalion of soldiers to Syracuse ordering Denys to release Plato immediately. Fearful of a war, Denys agreed and Plato was released.

Mr Ruche read back over his notes, then taking his pen wrote:

> The world of mathematics was greatly expanded by the Pythagoreans. They brought to it disciplines such as mechanics. Their often mystical view of numbers did not stop them from establishing arithmetic as a science. The first proofs in the history of maths were developed by them. They proved that the square root of 2 was an irrational number. It had been irrational numbers that had proved Hippasus' downfall. In geometry, they demonstrated that the sum of the angles in any triangle is 180°.

Mr Ruche was satisfied, he had more than enough for his next session on Pythagoras & Co. He put away his notebook and wheeled himself towards the door of the Rainforest Library.

Two days later, after much secretive preparatory work by Max and Mr Ruche, Jonathan and Lea came into the screening room – the empty studio beside the library – for their presentation on the school of Pythagoras. The room was dark; they could just make out some chairs and Albert, sitting in the far corner.

After a long silence, a light glowed faintly through the curtain which divided the studio. There was a tinkling, like musical chimes. On the other side of the curtain Max had placed four identical vases on a small table. The first was empty, the second half-filled with water, the third one-quarter filled, the fourth filled to one-third. Max held two small hammers. He tapped gently on the empty vase, then on the vase which was half-full. Two distinct notes rang out. Then he tapped the two together, making a more harmonious sound than either alone.

'An octave!' squawked Sidney.

There was a moment's silence, before Max tapped the empty vase and the one that was one-third full. They rang out.

'A fifth!' said Sidney.

Another silence, then Max tapped the empty vase again, this time with the one that was one-quarter full.

'A fourth!' said Sidney.

Max could barely make out the sounds himself, but he was determined to carry out the experiment. Jonathan and Lea listened without really understanding what was going on. Albert listened too, but didn't try to understand. Mr Ruche wondered whether taut string would have produced more distinctive notes. He worried that he had chosen the more spectacular experiment, rather than the most useful, but it was too late now.

'Pythagoras saw numbers ever...' The parrot's voice trailed off, and Jon-and-Lea heard a fluttering of wings before he cleared his throat and began again:

'Pythagoras saw numbers everywhere. Everything that is can be enumerated. He first found numbers in music.' The bird's voice trailed off again and Mr Ruche took over:

'Using a simple device, Pythagoras had made an extraordinary discovery: that the difference in musical notes is a difference between two numbers. The octave you heard is produced by the two vases, one empty, one half-filled; it is defined as a ratio of one-half; a musical fifth is two-thirds and a musical fourth is three-quarters. Can you think of a series of numbers simpler than that?'

Mr Ruche went on, 'So the relationship between numbers could be used to describe musical harmonies. In fact, harmony is simply a series of sounds based on numerical ratios. The notes are numbers, and the music is mathematics.'

From behind the curtain, a woman's voice began to sing Bach's *Ich habe Genug* a cappella. As the soprano's voice swooped lower and lower, Mr Ruche continued. 'Pythagoreans believed that the whole universe was a series of harmonies, that the heavens themselves were regulated by a musical scale. They called it the Music of the Spheres. Pythagoras invented a word for it: the *cosmos*. Order and beauty: to him the history of the universe was the struggle between cosmos and chaos.'

Mr Ruche looked down at the text he had prepared. 'Those three notes – mathematical harmonies – sounded out for Pythagoras the first mathematical law of nature. After that, he began to find numbers everywhere.

'Pythagoras and his school were determined to discover the mathematical laws inherent in the way nature regulated itself. To do this, they had to study numbers themselves. This was the beginning of *arithmetic* – the science of numbers, which they distinguished from *logistic* – pure calculation. In separating these disciplines, they elevated arithmetic above mere counting.'

Mr Ruche switched on the tape recorder and a voice boomed out through the loudspeaker: 'Attention! Attention! Listeners are asked to move to the other side of the curtain. The other side of the curtain.' Jon-and-Lea stood up (noting that they were listeners rather than viewers) lifted the curtain and went through. Here, three spotlights glowed in the darkness. One illuminated Max, who was sitting at a low table on which a number of objects – including the four musical vases – were arranged. The second spotlight was on Sidney, on his perch in front of a partition. The third and most powerful spotlight shone on Mr Ruche, who was on a small podium, surrounded by records and tapes and a hi-fi system with two powerful speakers. Mr Ruche picked up a sheet of paper and began to read:

'Pythagoras began by creating a catalogue of numbers, beginning with 1 – which seems so natural to us now that we assume it has always existed. He divided whole numbers into odd and even numbers: those which are divisible by 2 and those which aren't. He then went on to establish the rules of calculation...'

Sidney chipped in, 'Even plus even equals even; odd plus odd equals even; odd plus even equals odd.'

Mr Ruche continued, '...and the rules of multiplication.'

The parrot chipped in again, 'even times even equals even, odd times odd equals odd; odd times even equals even.'

The door opened on the other side of the curtain and a fresh breeze blew into the studio. Perrette slipped quietly into the room just as Jon-and-Lea applauded. She spotted Albert and sat down near him.

The loudspeaker broke the silence again: 'Attention! Attention! This is a revelation, this is a revel...'

Mr Ruche cut the sound and picked up the story. 'I have an important announcement: Pythagoras' theorem is not, I repeat not, by Pythagoras.' There was a scattering of applause at this revelation. Lea didn't quite know why she was so impressed; Jonathan was indifferent.

'Credit where credit is due', Mr Ruche continued. 'Before Pythagoras was born, the Egyptians and the Babylonians had discovered a link between triple numbers. That was the link made famous by Pythagoras. A collector named Plimpton acquired a Babylonian tablet on which a scribe had engraved a dozen triplets, indicating definitively that the sum of the squares of the first two was equal to the square of the third.'

Mr Ruche signalled to Sidney. The parrot arched himself on his perch. Max stood up.

'Three blocks of wood', Sidney announced. Max picked up three pieces of wood from the table. Sidney went on, 'The first is 3 units long, the second 4 and the third 5.' With his hands, Max indicated three units, then four, then five.

'Looks like they've been rehearsing!' moaned Lea.

'Yeah', said Jonathan. 'Max looks like a flight attendant.'

It was certainly true that Max had a fixed smile, and his mechanical hand gestures did look a little like a steward showing passengers how to put on their life jackets.

The parrot continued, '3 squared is 9; 4 squared is 16, added together they make 25, which is 5 squared. A triangle with sides of these lengths is a right-angled triangle.'

As Sidney spoke, Max traced the equation $3^2 + 4^2 = 5^2$, then he placed the three pieces of wood together to form a right-angled triangle:

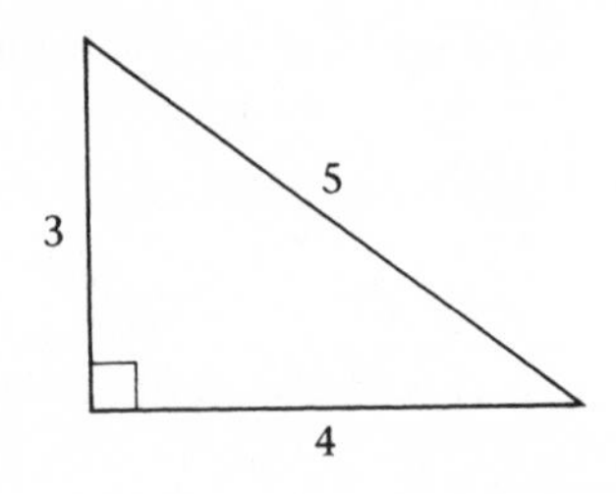

'So, what does the theorem tell us?' asked Mr Ruche. 'It tells us that there is a relationship between the lengths of the sides of a right-angled triangle which can be expressed as $a^2 + b^2 = c^2$.' Mr Ruche poured himself a glass of water and took a sip.

On the other side of the curtain, Perrette stretched out her legs and kicked off her shoes. She was tired from her long day in the bookshop. She could hear everything, but couldn't see what was going on. More importantly, she thought, she couldn't see what all this had to do with Grosrouvre's letter and the questions it posed. Jonathan, meanwhile, interrupted Mr Ruche: 'I'm not trying to defend Pythagoras, but...' (in fact, that was precisely what he wanted to do: for some reason, perhaps his long hair and his Grecian nose, he identified with this Greek traveller) '...but you said yourself that there's a difference between giving examples of something and proving it. The Egyptians and the Babylonians gave examples, but were they able to prove the theorem?'

'Apparently not', admitted Mr Ruche.

'In that case it is Pythagoras' theorem. Credit where credit's due!' said Jonathan triumphantly.

Lea asked Mr Ruche, 'So what was the curtain for?'

'I'm surprised you didn't ask before. I'm a bit worried that you're learning to be patient', Mr Ruche said sarcastically. 'The curtain was there so that you would know how it felt to be an apprentice applying to join Pythagoras' school. Pythagoras was rigorous in his selection. He began by seeing whether the disciple could "hold his tongue" – his words, not mine. Could he remain silent during lessons and, more importantly, keep secret what he learned? The first test depended not on what the disciple said, but on what he did not say. The schoolroom was divided into two by a curtain. Pythagoras would sit on one side and the candidates on the other.

They could listen, but could not see. This period lasted for five years.'

Lea was outraged: 'Five years! Don't look, just shut up and listen – it sounds like a cult.'

Max was fuming too. 'What about deaf people?' he thought. 'Aren't we entitled to learn? How could we be expected to find out anything if we were behind a curtain? I know I wouldn't stand for it.'

Mr Ruche could tell what was bothering him and said apologetically, 'Sorry, Max, that's the way it was.' Then he continued. 'The curtain played an important role in the Pythagorean school. To go beyond it meant that one had passed the tests. Those on the far side of the curtain were called *exoterics* and those who sat with Pythagoras were *esoterics*. They alone could hear and see him.'

'So when you brought us behind the curtain it was because you thought we were worthy of being esoterics, is that it?' Jon-and-Lea asked in unison.

'Precisely', said Mr Ruche.

'Why?'

'Because you managed to keep your mouths shut the whole time you were on the other side. I must admit I was surprised, but you both managed to hold your tongues.'

'So it was a trick?' Lea gave Jonathan a knowing look.

'Not a trick, a test', Mr Ruche corrected.

'And if we hadn't kept our mouths shut?'

'You would have stayed on the other side of the curtain. Max and I decided that was fair, and Sidney agreed.'

Hearing his name, the parrot took off from his perch and flew around the room. He flew straight into the curtain. Max tried to open it, but only succeeded in pulling the whole contraption down on top of himself. It was only then that they noticed Perrette sitting quietly on the other side. No one had heard her come in.

'How long have you been there, Mum?'

'Since the bit about Pythagoras' theorem', she smiled at him.

Albert shifted in his seat. He was sound asleep. They all burst out laughing, but even that didn't wake him. Mr Ruche carried on, like an actor trying to remain calm as the scenery falls down around him.

'The writings of Pythagoras' school were also secret and were deliberately written to have a double meaning: one which anyone could understand, but the other comprehensible only to the initiated. Pythagoreans often spoke of *sumbola* and *ainigmata*: symbols and enigmas.' As he said this, he thought of Grosrouvre's letter which, he was sure, was a real Pythagorean text, full of symbols and codes.

'Most of their knowledge was passed on by word of mouth. This further divided the school into *acousmaticians*, who were told the results of a proof, but not the proof itself, and *mathematicians*, who knew both the proofs and the results.' Mr Ruche wondered aloud who Grosrouvre's loyal friend was – the person to whom he confided his proofs after he had burned them. Like Pythagoras' disciples, his friend would have had to learn the proofs by heart, though he did not need to understand them – only to be able to repeat them. The friend did not have to be a mathematician, just an acousmatician, as Pythagoras called them.

There was silence for a moment. Then Lea smiled and said, '"In search of a Rainforest Acousmatician." That sounds like a good title for our end-of-term paper.'

'What about us? Are we acousmaticians or mathematicians?' asked Jonathan.

'That depends on how well you understand the proofs. Only time will tell.'

'Every pupil of Pythagoras had to train his memory', Mr Ruche picked up where he had left off, 'Every morning, before getting up, he had to remember the precise events of the day before – everything he had seen or said or done and everyone he had met.'

'What happened to the ones who weren't accepted?' Lea asked.

'When a disciple first put himself forward to the school, he had to give all his worldly goods to the community.'

'Like I said, it sounds just like some sect', said Lea.

'With one condition,' Mr Ruche went on, 'that anyone who was not accepted be given twice the value of the goods he had brought.'

'You mean the stupid ones left with more money than they came with?' Jonathan asked incredulously. 'You don't find religious sects doing that nowadays. They're more likely to bleed you dry!'

'He was rewarded financially for what he was not able to gain in knowledge', added Mr Ruche. 'But...' (he left a pause for emphasis) '...as soon as it was decided that he should leave, a grave was dug for him.'

'They didn't kill him, did they?' shouted Max.

'It was supposed to be symbolic, Max', said Lea.

Perrette looked up, her eyes shining. 'The death was symbolic, but the grave was real enough. If someone had come along and seen it, they would have assumed that he was really dead. So it is possible to believe you have proof that someone is dead, even though he's still alive.'

'What is she going on about?' Lea wondered.

Max moved closer to his mother, and all the others listened carefully.

'You're talking about Grosrouvre, aren't you?' asked Mr Ruche. 'But they found a...' (he couldn't bring himself to say 'a corpse') '...they found his body. It's one thing to find a grave, but I think you're confusing a tomb with a body.'

'I'm not confusing anything', said Perrette.

'So what are you saying?' Mr Ruche sounded angry.

'What I'm saying is, how do we know that the body in the ruins of your friend's house was really his?'

This possibility hadn't occurred to them before now. Mr Ruche turned to her: 'Perrette, the commissioner told us in the letter that it was Grosrouvre's body.'

'I don't understand you, Mr Ruche. Do you want your friend to be alive or don't you?'

'What do you mean, "Do I want him to be alive?" What I want doesn't make any difference. It won't bring him back to life.'

'And if you don't have any proof that he is dead there's no reason to kill him off.' Perrette was angry now too.

'Just a minute...what do you mean, kill him? Are you saying I killed Grosrouvre?'

'Calm down a minute. I'm not saying that at all. I'm just saying that we don't have any proof that he's dead.'

'No proof?' Mr Ruche was furious now. 'What about the charred body they found in the ruins – isn't that proof enough?'

'No. All the body proves is that the person in the house when it burned down is dead. It doesn't tell us who that is or even whether they died in the fire. I mean, did anyone identify the body? Was there an autopsy?'

'May I remind you Perrette, that you're the one who suggested Grosrouvre was murdered.'

'I'm not trying to contradict what I said before. Let's just consider every possibility.'

'Is anyone hungry?' Lea interrupted.

'But if it isn't Grosrouvre, then who is it?' asked Mr Ruche.

'Let's just try to find out first whether it is Grosrouvre', said Perrette.

'Well, you lot may not be hungry, but I'm starving,' said Lea.

'OK, OK, let's leave it there', said Mr Ruche. 'We can talk again after we've eaten. We can do a...what is it you kids call it?'

'An all-nighter?'

'That's it. If we have to, we can do an all-nighter.'

At this, Albert woke up. His cap had slipped off, but his cigarette was still stuck to his lower lip.

'I think I might have dropped off', he said. 'I did an all-nighter last night out at the airport. It's good money, airport runs, but it takes it out of you.'

'What about Albert, Mr Ruche?' asked Lea. 'Is he an esoteric too? He didn't say anything and you said that was the rule.'

'Albert,' declared Mr Ruche, 'you have just been made an esoteric. You now belong to the ranks of the Pythagoreans.'

'I do not. I don't belong to nothing. I've always been a bit of a loner. You won't catch me joining associations or parties or unions. I wouldn't join a bowling team, me!'

CHAPTER 8

Just a Fraction

Irrational numbers; imaginary journeys & proving the impossible

Mr Ruche manoeuvred his wheelchair onto the lift, pressed the button and rose up above the courtyard. The session about Pythagoras had been long and tiring. He was beginning to regret promising to do an 'all-nighter'. The lift squeaked like a roller coaster at a funfair, just before it plunged downhill and everyone screamed.

Max had stayed in the screening room. He didn't notice Perrette, sitting in the dark, at the end of the room. She was wondering why had she been so short with Mr Ruche. She was surprised to find herself worked up about the death of a man she had never met, someone she hadn't even heard of until recently. Things had changed since Mr Ruche received Grosrouvre's first letter, she thought. Before that, they had all lived together peacefully. They rarely argued. Now they were caught up in a strange mystery involving death, books, maths, fire. She didn't know whether it was a blessing or a curse. All she knew was that for the first time they had all come together as a team; even the parrot seemed to play its part.

While Max carefully folded away the curtain, Sidney flew down to the little table. He was thirsty and poked his beak into one of the vases, but he couldn't reach the water. He tried two others, but still had no luck. Max saw him and came to his rescue. Perrette watched them, amused. Max took the vase marked '$\frac{1}{3}$' and poured it into the one marked '$\frac{1}{2}$'. Sidney leaned in, but the water was still too far away. Max picked up the vase marked '$\frac{1}{4}$' and was about to pour it in when Perrette noticed Mr Ruche's notebook close by on the table. 'Max, stop!' she said, running forward. It was too late – he had already poured the water from one vase into the other. The water spilled out over the rim and onto the notebook. Max had sensed Perrette rather than heard her. He dried the notebook on his shirt.

'How did you know it was going to overflow?' he asked.

'I added it up and realized it would.'

Perrette had been working on the till in the bookshop for ten years now and as she keyed in the prices, she liked to compete to see which of them could calculate more quickly. Man against machine – well, woman against machine. She felt a little like a chess-player facing a computer.

'How did you know?'

'When you poured the water from these two vases into that one, you added their contents together: $\frac{1}{2}+\frac{1}{3}+\frac{1}{4}$ comes to $\frac{13}{12}$. Since $\frac{13}{12}$ is more than 1, then the contents were greater than one whole vase and it *had* to overflow. By my calculation, that makes $\frac{1}{12}$ of a vase of water all over Mr Ruche's notebook. I don't think he'll be very happy.'

Perrette inspected the notebook. The pages were wet but the text was still legible.

Max was impressed.

'You added all that up in your head? Cool, Mum!'

Perrette blushed.

'No, really, Mum, You're pretty good.'

In the small kitchen, the coffee was brewing slowly in the pot. Albert waited for a moment before pouring himself a large cup. He drained the cup, and then had a second coffee so that he stood some chance of staying awake through the 'all-nighter'.

'Why do you work nights if it tires you out so much? Is it the money?' asked Jonathan.

'Sometimes,' said Albert, 'but last night I worked because I wanted to go to Rio.'

'Rio?'

The knife that Jonathan was using to cut some slices of ham slipped and hit the chopping board. Jonathan liked his ham sliced as thinly as possible.

'When I'm tired of Paris or if I feel like getting away from it all, I go to the airport: Orly or Roissy, it doesn't matter. Yesterday, when I woke up, I thought, "I feel like going to Rio!" I checked the airline timetable – I always have a copy beside my bed – and worked out that I'd have to get to Roissy at five in the morning. I arrived just as the plane was landing and picked up a Brazilian couple. When they got in the cab, I said to them, I said, "So, what's new in Rio, then?" I talked to them about Rio things I'd heard from a guy I'd had in the cab a couple of weeks ago before. The woman said, "You certainly know Rio very well. When were you there?" And I said to her, "I've never set foot there in my life". You should have seen the look on her face.'

'I always ask passengers about the cities they come from. I like to find out their favourite restaurants, the parks they like to sit in, that sort of thing. I might not have travelled much, but I know lots of cities I've never been to like the back of my hand: New York, Tokyo, Bogota, Singapore. I never read the guidebooks – that would be cheating.' In fact, Albert had read one guidebook about Syracuse. It was a city he had always wanted to visit, but there were no direct flights to Paris and no passengers to tell him what it was like.

'Cities, mind you, not countries. Countries only exist on maps, but cities...cities are real places.' His passion for airports, he confided to Jonathan, had come from the only trip he had ever taken – to Rome, many years ago. On the first day he'd lost his papers and his plane tickets, caught flu and had to stay holed up in a hotel for the whole of his visit.

'You don't know Manaus, do you?' Jonathan interrupted.

'No, where is it?'

'In Brazil – in the Amazon.'

'The only places I know in Brazil are Rio and Brasilia. 'Fraid Manaus doesn't fly long-haul to Paris.'

Albert finished setting the table just as Perrette and Max walked into the dining room with Sidney, and Lea came downstairs. They sat down to dinner. The ham and salad were excellent, and when they had finished they went back to the screening room. Mr Ruche was tired, so Perrette suggested that they leave the meeting until the morning, but he insisted. Perrette helped him lift the wheelchair onto the podium. Albert sat in the front row – in the stalls – determined to stay up all night. Sidney had stayed behind on his perch in the dining room, tired from the long session that afternoon.

'Since some of you can't wait to find out about how irrational numbers rocked the world more than two thousand years ago, we've organized this night school', began Mr Ruche. 'It was in the fifth century BC, in a town called Crotona somewhere in the Greek Empire – probably what we now call southern Italy, and it's a drama in three acts. Act 1: Everything is a number. Act 2: If a square has sides one unit in length, the diagonal of the square cannot be expressed as a number. Act 3: Therefore measurements must exist which cannot be expressed as numbers. This idea, which the Pythagoreans put forward themselves, suddenly placed their view of the world in jeopardy. It was essential that it be kept secret.

'Let's start with Act 1: Everything is a number. These numbers, which Pythagoreans believed described the world, the cosmos and its harmony, were whole numbers and fractions – which are simply divisions of whole numbers. In fact, they did not include all whole numbers, just positive numbers, because negative numbers were unknown in the ancient world.'

There was a murmur of surprise from the audience. Mr Ruche continued:

'Greeks allowed for all fractions, but in Egypt, there was only $\frac{1}{2}$ and a number of other specific fractions. There was no $\frac{22}{7}$, for example. The basic principle behind what were later called *rational numbers* was that they expressed geometric sizes: they were measurements.

'Act 2: the problem of measuring the diagonal of a square with sides 1 unit in length.' Taking a piece of paper, Mr Ruche drew a square cut by a diagonal and held it up for all to see:

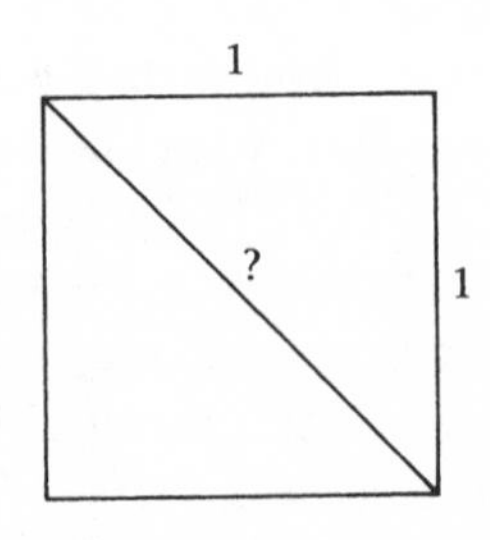

'What is the relationship between the sides of a square and its diagonal? Let's take the simplest square, with sides which are 1 unit long. If we draw a diagonal through the square, we get two equal isosceles triangles. The hypotenuse of the triangles is the diagonal of the square. So, what does Pythagoras' theorem tell us?'

The question was intended to be rhetorical, but the whole audience erupted with the answer: 'The square on the hypotenuse is equal to the sum of the squares on the other two sides.'

'And if we remember that the square of 1 equals 1,' went on Mr Ruche, 'then the equation we need to solve to work out the hypotenuse is "Square of the diagonal is $1^2 + 1^2 = 2$." So, the solution is that the diagonal is equal to the square root of 2!'

Mr Ruche wheeled himself to the edge of the podium, closer to his audience. He looked along the upturned faces and asked dramatically, 'And what is that number? The Greeks tried to find it, but they couldn't find any number – whole or fraction – which when squared gave 2. They therefore had to ask the question, "Does such a number exist?" And if it doesn't, how to prove that it doesn't? To prove that something exists is easy, what's difficult is to prove that something does not exist. The only thing they could do was prove that it *could* not exist.

'They decided to move from the uncertainty of finding it to the certainty that it could not be found. And that's what the Pythagoreans did: they proved that a rational number whose square is 2 cannot exist. If the sides of the square are 1 unit long, then there is no number which can represent the diagonal. There is no common measure between the sides and the diagonal.'

He held up the piece of paper again, though not as high as before. He was very tired now. Perrette thought they should stop, but she knew that on no account could she interrupt him now.

'Look at the diagram,. Does it look as if there is no common measure between the sides and the diagonal? No. There's nothing that suggests it cannot be measured. The incommensurable cannot be seen – it has to be proved by thought and hard work.

'Act 3: How did Greek society react to this revelation? This simple diagram represents a black hole which threatened to swallow up all the certainties on which Greek society was based. The fundamental link between numbers and space, which mapped out the universe of Pythagoras, was in ruins. And all this had originated from a square, one of the two basic shapes in ancient maths. Worse still, it had all happened using Pythagoras' theorem and the division of numbers into odd and even – both discovered by the Pythagoreans themselves.

'So, what did they mean by "incommensurable"? They meant that no common measure could be found between the two. If one number represents the first, there is no number to represent the second, which means that it is impossible to measure them against each other – even though, when you look at them, they both seem to be just as...' (he searched for a word) '...just as real. The coexistence of these two lengths proved that life was more complicated than numbers.

'I have just drawn this diagonal, but I can't measure it. Before this, anything anyone had constructed could be measured. Now, the link between what could be made and what could be measured was broken. The revelation was this: that there are certain measures for which there was no number, so Pythagoreans called them *alogon*, "inexpressible".'

Mr Ruche was exhausted, but it was obvious that he was happy too. This was real philosophy. It had been forty years since he had felt like this. This was his best performance – a one-man show, without Max, or the loudspeaker, or even Sidney. Perrette was impressed, but she was worried that he was overdoing it.

Mr Ruche went on. 'This was the "mathematical scandal" that Hippasus had disclosed to the outside world. For this, he died in a shipwreck. The shipwreck was, in a way, related to the harmony and the order of all things in the world. It had been provoked, they believed, by this proof which upset the sense of harmony. The first mathematical proof was a proof of something impossible.'

'It can't have been easy to prove', said Perrette, thinking aloud.

'Actually, Perrette, given the importance and the consequences of the proof, it was not too difficult.' Mr Ruche fell silent. He was too tired to explain the proof to them now. The meeting broke for the night.

Jonathan lay on his bed in the attic and listened to the tap running. 'To move from uncertainty to certainty' – Mr Ruche's phrase bounced around in Jonathan's head like a ball on a snooker table.

Lea came out of the bathroom, her hair wet, and went into her room. She propped a mirror in the folds of the blankets and started to colour the fringe blue. 'The incommensurable cannot be seen.' Jonathan stared at his sister for a long time.

'We should get to work on that proof', said Jonathan suddenly, still watching her.

Lea stopped what she was doing. 'You were spying on me.'

'I want us to try and do the proof that Mr Ruche didn't explain.'

'Why?'

'Because I'd like us to move from uncertainty to certainty. We know so little about ourselves, it might be nice to find certainty even if it's only in maths.' Lea nodded, and put down her hairbrush. They gathered the maths books onto Jonathan's bed and worked as they had never done in their lives.

After some time, they realized that everything hinged on the fact that Pythagoras had divided numbers into odd and even. The principles he had devised proved that a number could not be both odd and even. This was what they used to prove Mr Ruche's theorem. They finished just before dawn, and went to bed, sleeping soundly until noon and missing a morning's school.

They walked into the dining room where Perrette and Mr Ruche were finishing lunch. Mr Ruche had dipped a pear into his glass of wine, while Perrette was cutting a small goat's cheese into thin slices. Jonathan began.

'Last night, Mr Ruche said that proving that the square root of 2 is an irrational number was easy. He said he was too tired to explain, but we think he chickened out.'

'I did not chicken out', choked Mr Ruche, spilling wine on his white shirt.

Lea took the little blackboard Max had used at primary school and announced, 'A proof that the square root of 2 is not a rational number.'

'Let's say,' began Jonathan, 'that the fraction $\frac{a}{b}$, when squared, equals 2.'

'Meaning,' picked up Lea, writing the equation on the blackboard, 'that $\frac{a^2}{b^2} = 2$.'

'Let's assume that it is a simple fraction and that a and b are co-prime, meaning that there is no number which can divide into both.'

'This would mean that a and b cannot both be even numbers', said Lea confidently. Now, if $\frac{a^2}{b^2} = 2$, then it follows that $a^2 = 2b^2$. This means that a must be a multiple of 2 – and therefore must be an even number.'

Perrette stared at the twins in amazement.

'Because,' said Jonathan, glancing at his mother, 'a square is only even if the root is even.'

'Like I said,' repeated Lea, 'a is an even number. a is a multiple of 2 – let's say, then, that $a = 2c$.' Jonathan wrote the equation on the board.

'Slow down, slow down...' said Mr Ruche, trying to keep up.

'So, if we take our initial equation $a^2 = 2b^2$ and replace a with $2c$, we get $(2c)^2 = 2b^2$. This means $4c^2 = 2b^2$, so $2c^2 = b^2$.'

'Therefore b is a multiple of 2...'

'I know my eyesight isn't what it was, but your writing is terrible', said Mr Ruche.

'Therefore,' announced Jonathan, 'if b^2 is a multiple of 2, b^2 is an even number.'

'Which, as I pointed out earlier, means that b is also even', said Lea.

'So, there are two incontrovertible statements: first, a and b cannot both be even numbers; secondly a and b *are* both even numbers. Of course that is impossible. Why do we have a result which is patently absurd?' Jonathan asked Lea with the menace of a prosecuting lawyer.

Perrette and Mr Ruche looked at each other, as if to say, 'Do you see what I see?' The twins were getting worked up about maths. Max looked at their astonishment happily; he was proud of the twins.

'Well, why *do* we have a result which is patently absurd?' Jonathan asked again.

'Because of my hypothesis', Lea looked down.

'And what was your false hypothesis again?'

'That there is a fraction which, squared, equals 2: $\frac{a^2}{b^2} = 2$.'

'Then get rid of it!'

The twins picked up two forks from the table and began tapping a reggae beat on two wine glasses, singing:

We know that
By our action
We've proved there is no fraction,
Which, squared, could equal 2,
It's true, it's true.

A round of applause greeted this Bob Marley version of their proof. They went over to Mr Ruche. 'So, are we acousmaticians or mathematicians, then?'

'Let's see: memory for facts, good; understanding of proofs, good.' He tapped the table. 'I think we can safely say that you are mathematicians!'

CHAPTER 9

Night Boat to Alexandria

Black tie; white sands & an introduction to Euclid's Elements

It was late November, three months since Grosrouvre's letter had shaken up the household on the rue Ravignan. He would have been proud of the revolution he had started.

Perrette and Mr Ruche had finished cataloguing the Rainforest Library, but since the late-night meeting they had made little headway. Though they had worked hard, Mr Ruche was forced to admit that their work lacked rigour. They had not been at all meticulous. It was time to get back to work. He called a meeting to discuss this, and Max, who had noticed that Sidney was more inclined to talk in the evening, suggested that the session should be held at night. This was no ordinary session, and Mr Ruche decided that their new rigour called for a more formal occasion. He issued a beautifully printed invitation to 'A discussion to be held in the Rainforest Library at 8 p.m. sharp. Dress: formal.'

Jon-and-Lea arrived at the door of the Rainforest Library in full evening dress – well, in something that passed for evening dress. Lea had borrowed a long dress with a slit up the side from a friend, and a velvet stole from Perrette. She was tottering slightly on her high high-heels. Jonathan looked more like a dandy than a duke. He had found a gold tie that set off his black shirt beautifully. He was wearing a silver-grey jacket two sizes too small and trousers that defied description, but were neatly creased. His footwear, however, was incongruous: he was wearing sandals.

As the usher, Max greeted them at the door, took their invitations and led them to the row of velvet armchairs where they were to sit. Then the room was plunged into darkness. A spotlight flared out from the centre of the room and began to revolve slowly, lighting up the bookcases like a lighthouse. The spotlight moved around the room and then flooded the great bay window that looked out onto the courtyard. Jon-and-Lea could hear the gentle sound of waves breaking. All they needed now was the smell of the salty breeze and the sound of gulls, and they could be by the sea. The light began to dim, and a voice issued from the loudspeaker:

'Attention! Attention! You have just entered the Great Library of Alexandria. We kindly request that visitors do not use flash photography. Smoking and chewing gum are strictly forbidden.'

Lea kicked off her shoes and nudged them under her armchair with her foot.

Mr Ruche began: 'Neither Thales nor Pythagoras visited the Great Library of Alexandria when they went to Egypt, for the simple reason that Alexandria had yet to be built. It was founded some three hundred years after their deaths by Alexander the Great, in 331 BC, after his conquest of Egypt. The town sprawled along a strip of land between the sea and Lake Mariotis, and was surrounded on all sides by sand and swamp. Near the gates of the city lay a small island, defending the city against the constant assault of the sea: Pharos. In honour of Alexander, the architect designed the city to represent a chlamys – the heavy purple cape of the Macedonian soldiers who accompanied the general. It was a perfect rectangle, a geometric city, with a grid of streets which met at right angles.

'The city had a population of three hundred thousand, not including slaves. Unlike Athens it was cosmopolitan: its inhabitants included Egyptians, of course; Greeks who had crossed the Mediterranean to seek their fortune; Jews from neighbouring Palestine; and mercenaries from the four corners of Europe who had come to sign up in King Ptolemy I's army: Scythians, Thracians and the feared Gauls.

'Disembarking from their ships, visitors found a vast luxurious metropolis, traversed by canals, paved with flagstones, the streets wide enough to drive four chariots abreast. Long lines of tall marble columns seemed to join heaven and earth, each column topped by a marble capital so large that a hundred men were needed to lift it. The buildings, decorated with coloured stone and marble, were safe from the threat of fire.

'The city was a constant hive of activity. It had two ports: one in the east, the other in the west, so that regardless of the prevailing wind a ship could dock in safety. Ships came and went day and night from all parts of the Mediterranean – Alexandria was the market place of the world. Huge stores selling cereals and every kind of goods stretched for miles along the quays. This was the gateway between Europe and Africa, between Greece and Egypt, where the Gods of Olympus met Isis and Osiris. It was the repository of Greek learning for seven centuries.'

Jon-and-Lea could easily imagine the city. They would have given anything to be in the warm, white marble streets of Alexandria. Instead, they were stuck in Paris, where the skies were grey and it rained all the time. But in their heads they were already planning a much greater journey for the summer. It was a secret which they talked about only late at night, under their skylights.

Mr Ruche went on: 'Cities everywhere were clamouring for status, but it was Alexandria that took the place of Athens. For seven hundred years Alexandria was the centre of intellectual thought in the Western world.'

A thousand miles away in wintry Paris, the spotlight flared up in the Rainforest Library, lighting in turn the four corners of the room. That was the signal for Sidney to speak:

'I ask of the rulers and the governments of all the peoples of the world that they send to the city of Alexandria copies of the work of their poets and writers, of their orators and sophists, their doctors and their soothsayers, their philosophers, their historians...'

'Who made this request?' asked Max, who was clearly the compere.

'Ptolemy I, known as Soter, "the saviour"' Mr Ruche was speaking now. 'The first of the Lagides dynasty. He was a friend of Alexander and came to the throne after the death of the general. His edict was taken by messengers across the vast empire that Alexander had founded, but was now being torn apart by pretenders vying to succeed him.

'A man named Demetrius penned the message. He had been a successful politician and philosopher in Athens but was forced to leave the country when the tide turned against him. He found refuge in Alexandria, where Ptolemy welcomed him.'

Mr Ruche dropped his voice almost to a whisper. 'Demetrius wanted to create institutions which would bring to life the dreams of Aristotle for a centre of universal knowledge. There had been great schools before this. Plato's Academy still stood in the grove of Academus, in the heart of Athens. Theophrastus, a disciple of Aristotle, had founded the Lyceum, whose pupils would discuss philosophy in the shaded cloisters of the outdoor gymnasium. Demetrius decided that he would found an institute greater than these: the Great Library and the Museum, which together would encompass all the world's knowledge. Ptolemy was immediately taken by the idea.

'Nothing like this had ever been attempted, and it was a complete success. Wise men flocked to the Museum and the Library, whose vast buildings were contained within the Museum, filled with manuscripts. But there was another building in Alexandria which rivalled these: the great lighthouse – one of the Seven Wonders of the World.

'It stood so high in the night sky that sailors might have thought it was a new star. It was built on foundations impervious to the sea. At ground level was a square tower over two hundred feet high, on which stood a second tower, this one octagonal, almost twice the height of the first and on this, in turn, was a third tower. This last tower was a slender cylinder thirty feet high, fashioned in white marble. At the top was a dome supported by pillars and under it a fire raged, its light multiplied by a battery of mirrors. While still on the high sea, some 30 miles out, sailors in the darkness were drawn like moths to the glow of the lighthouse leading them to the port.

'For sixteen hundred years the lighthouse of Pharos would light up the night sky around Alexandria.' Mr Ruche looked down at his notes. 'Then, in 1302, it was destroyed by an earthquake which shook apart the slabs of marble and tossed them into the sea.'

'How on earth did they build a lighthouse like that back then?' asked Jonathan.

'Huge buildings were a bit of an Egyptian speciality', said Lea. 'What I want to know is whether building the lighthouse cost as many lives as building the Great Pyramid of Cheops?'

'Which would you prefer, to be flattened while helping to build

Cheops or tossed into the sea by an earthquake at Alexandria?' asked Jonathan.

Mr Ruche ignored them and continued. 'The lighthouse lit up the night sky for the sailors, the Library lit up the human mind for anyone who was hungry for knowledge. Above the door of Plato's Academy were the words, "None may pass who is not a Geometrist". The Museum had nothing of the sort: it was dedicated to the muses, to each and all of them. While the Academy and the Lyceum were private schools, with income from their members, the Museum was a public institution which survived on a generous subsidy from the Pharaoh.

'The Museum stood in the centre of the city, near the palace, not far from Ptolemy's private port. It was built in the purest Hellenic style, with shaded courtyards and ample gardens. Dotted about were reading rooms which were light, calm and spacious. There were rooms set aside for conversation, and others where people could rest. Paths lined with columns and fountains ran along green spaces where exotic animals roamed. There was a gallery for paintings and one for sculpture. Everything was designed to encourage work and thought. Plato's Academy had been attended by many great scholars: Theaetetus, Eudoxus and Archytas among them. The Museum of Alexandria counted Eratosthenes, Apollonius and the blind mathematician Dositheus among its scholars, but one of the first and certainly the most famous was Euclid.

'Those who worked at the Museum were hand-picked by the king himself. But, aside from the status conferred on the scholars in being invited to work at the Museum, there were many material benefits, including board and lodging and a tax-free income. The Library was also at their disposal day and night.

'To assemble any library is a mammoth task, but to fill the empty shelves of the colossal Library at Alexandria with manuscripts from across the known world was like one of the labours of Hercules. But the Library soon came to contain four hundred thousand rolls of papyrus. The authorities of Alexandria had sent out "book-hunters" to comb the markets of every town on the Mediterranean and bring back manuscripts at any price. And if they could not buy them, they would find other ways of acquiring them: theft, blackmail, assault.'

'Do you think Grosrouvre did things like that to assemble the Rainforest Library?' asked Max.

'Who knows?' In fact, from Grosrouvre's letters, Mr Ruche knew all too well that not all the books had been honestly acquired. 'Every ship which arrived in Alexandria was boarded by soldiers who set about searching the baggage for manuscripts. Ptolemy had issued an edict that any manuscripts brought to Alexandria were to be confiscated and taken to the Great Library. They were carefully copied by the scribes, and the originals returned to their owners. If the manuscript was rare, then the copy was given to the owner and the original kept.'

'That's a swindle!' exclaimed Jonathan, 'I show up with a book – a collector's piece – and I'm sent away with a cheap photocopy. And I bet if I complain, I'll get banged up. I don't think much of this guy Ptolemy!'

'Well, that's the way it was', said Mr Ruche. 'They were hardly cheap copies though. Both the originals and the copies were written out on rolls of papyrus. The first manuscripts were kept in rolls – in Latin, *volumen*, hence the word "volume".'

'Where would you be without etymology?' Lea asked sarcastically.

'I think I might find words a little less interesting', said Mr Ruche. 'Each volume was made up of leaves of papyrus wound around a baton. The text was written in columns, in Greek or in Demotic, the standard written language in Ancient Egypt. To read the volumes you needed both hands – one to hold the end of the roll and the other to wind it out. The rolls were labelled and stored in pigeonholes in the Library. They were arranged by subject – literature, philosophy, science, technology – and each subject was arranged alphabetically by author. Greek literature spanning three centuries was there: the complete works of Homer; tragedies by Aescylus, Sophocles and Euripides; comedies by Aristophanes, Hippocrates' *Elements*, the work of Theaetetus and Theodorus, and Aristole's library, which Ptolemy had acquired with a lot of money and a lot of muscle.

'By the time the library had been filled, Demetrius was not there to celebrate. He had used his political wiles to favour one of Soter's sons and was condemned to death. In the end, he preferred to take his own life. He was the last of the great Athenians.

Ptolemy II took the name Philadelphus, meaning "he who loves his sister", and in the Egyptian tradition he married his sister Arsinoë. It is said she was a great beauty.'

Jonathan wolf-whistled.

'Philadelphus was also very beautiful, and apparently had magnificent golden curls.'

Lea whistled.

Mr Ruche's tone changed and he pointed to each of them in turn: 'Do you remember asking me, Lea, when we were talking about Thales, if there were any shortcuts in maths? And Jonathan, you asked me what maths was for?' The twins sat up straight. 'Well, I'm happy to introduce you to Euclid, who has answers for you – and I think you'll like them.

'One day, Ptolemy was visiting the library. Looking along the shelves of manuscripts, his eye fell on the many volumes of the *Elements*, each in its case. Turning to Euclid, he asked if there was not an easier way to understand geometry. Euclid replied, "There is no 'royal road' to geometry." You had to have guts to say something like that to a king. On another occasion, Euclid was trying to teach a student – an intelligent boy – and the student asked what he would get out of it. Euclid called over a slave and said, "Give him three drachmas, since he feels he needs to get something for what he has learned."'

'Reading you loud and clear, Mr Ruche', said Jonathan. He turned to Lea. 'What dear Mr Ruche is trying to say is that if you're going to tackle maths, there's no point being in a hurry or looking for shortcuts.' Mr Ruche and Lea nodded.

'Exactly, Jonathan', said Mr Ruche, 'your...um...theorem holds true not just for maths, but for the arts and for knowledge in general.'

'And love', added Lea.

'Probably, probably. That reminds me of something I once heard Grosrouvre say to one of his mistresses. It was in a café near the Sorbonne where we always met. Grosrouvre arrived very late and she asked, "What kept you, dearest?" Grosrouvre replied, "I was finishing a maths problem." The girl shrugged and said, "I don't get it – how can you spend so much time on that stuff? What use is maths anyway?" Elgar looked her straight in the eye and said, "And love, darling, what use is that?" He never saw her again.'

'We're getting off the subject', said Jonathan. 'Are you saying we should learn maths for no reason?'

'And that we take the long way round while we're at it?' added Lea.

Mr Ruche wondered at their incessant demands, but secretly he was thrilled because it gave him the opportunity to say something he had been waiting to say for a long time. 'Young people: from Aristotle you will learn logic and from Euclid, rigour.'

The library was plunged into darkness, and Jon-and-Lea could hear people hard at work and the sound of furniture being shifted around. Then, silence. Jonathan sat up as the lights came on. Everything had changed.

Mr Ruche was sitting on a podium in the space between the bookshelves. Just in front of him were a number of lecterns arranged in a semicircle, on each a manuscript written in his own hand. Mr Ruche indicated the thirteen lecterns and announced, 'Euclid's *Elements,* consisting of thirteen books. The author numbered them from I to XIII to indicate that they were part of a whole and that they should be read in a particular order. There is an internal order to each book, and an external order to the whole. This hierarchy is the architecture of Euclid's monumental work.'

After the Bible, Euclid's *Elements* is the second most frequently published book in the world, with over eight hundred editions to date. The copy in the Rainforest Library was among the oldest: an Italian translation by Niccolò Tartaglia, published in Venice in 1543. It must have cost Grosrouvre a fortune.

Max and Sidney made their entrance. Max was wearing a dinner jacket with tails which he'd picked up at the flea market. Jon-and-Lea tittered, and even Mr Ruche had to stifle a giggle. Max made his way to the lectern on the left, Sidney perched on his shoulder.

Mr Ruche began, 'Like any good Greek mathematician, Euclid gives pride of place to geometry and makes it the prelude to his work. The first four books are dedicated to it. Euclid's plan is clear: first, two-dimensional geometry, followed by number theory and then geometry in three dimensions. His method is also clear: to define shapes, calculate their area – with the exception of the circle – and then create them as three-dimensional objects.'

Mr Ruche pointed to Max and Sidney standing by the lectern.

'The first lines of the *Elements*, like a play, describe the characters of the epic story of geometry, to be played out in thirteen acts. This is the role of Euclid's definitions.' He nodded to the soloists, and Max and Sidney began a long duet:

'A *point* is that which has no part', sang Sidney.

'A *line* is length without breadth', Max chimed in.

'A *surface* is that which has length and breadth only.'

'A *plane angle* is the inclination to each other of two lines in a plane which meet each other and do not lie in a straight line.' Max caught his breath. 'Among all lines, one is remarkable: the straight line.'

This last line was picked up by Sidney: 'A *straight line* is a line that lies evenly with the points on itself.'

Mr Ruche interrupted to explain that on a straight line, it is impossible to determine any individual point. 'In other words, a straight line behaves identically with each of the points.' He nodded again to the soloists.

'Among surfaces, one is remarkable in being a plane', sang Sidney.

'A *plane surface* is one which lies evenly with the straight lines on itself.'

There was a moment's silence before Mr Ruche said, '*Angles*!' He held his arm up, bending it at the elbow. 'The word comes from *ankon*, meaning "elbow".' He locked his arm at 90°. 'Among angles, one is remarkable: the right angle.'

Max made a cross with his arms, and picked up the story. 'When two lines intersect, they form four angles. If the angles are equal, then they are all right angles.'

Mr Ruche continued. 'Geometric figures. First, the circle, which has only one form. After the circle come rectilinear figures – shapes made up of more than two lines – if you have only two straight lines, you can't create a figure: three is the minimum. Three lines form a triangle. There are wide triangles which have one obtuse angle, and narrow triangles which have one acute angle. And then you have the special ones: isosceles, where two sides are equal; equilateral, where all sides are equal; and right-angled triangles. After this come quadrilateral figures – shapes which have four sides. Here pride of place goes to the square, which, like the circle, has only one form. If you know the length of one side, you know all there is to know. Then the rectangle, which requires two pieces of information – the length and the breadth. There are hundreds of thousands of others: the rhombus, the parallelogram, the trapezoid, and others which have nothing remarkable about them.'

'You mean the ones we never study in maths?' asked Jonathan.

'That's right', admitted Mr Ruche, 'and it's hardly surprising. What is there to say about any old four-sided figure?'

'That it has four sides, four angles, two diagonals', said Jonathan.

'And that the sum of its angles is equal to 360°', said Lea.

Max waved his arm. Mr Ruche had forgotten something. Max announced, 'If there are two lines on a plane and you follow them out to infinity both ways – which would be pretty hard and would take, well, an infinite amount of time – and they don't meet, then they are *parallel*!'

Mr Ruche laughed. He picked up the story again: 'Book I ends with a bang: Proposition 47 is Pythagoras' theorem itself!*

'On to Book II. Once the characters in geometry have been defined, Euclid shows us what we can do with them. He cuts an angle in half and finds its bisector; he cuts a segment into two equal parts, which means finding the mean. He calculates areas and establishes when two figures of the same form are equal in area.

'To conclude his examination of plane geometry, Euclid presents the construction of regular polygons. In Books III and IV he deals with circles, and for each regular polygon he determines the *inscribed circle* and the *circumscribing circle*. A circle which circumscribes a polygon is outside the figure, but passes through each of the points; a circle which inscribes a figure is inside the figure, but touches tangentially each of the sides. The first of the regular polygons is the equilateral triangle, which would look something like this.' A slide appeared on the screen:

*A proposition is a formal statement of a theorem, together with its proof.

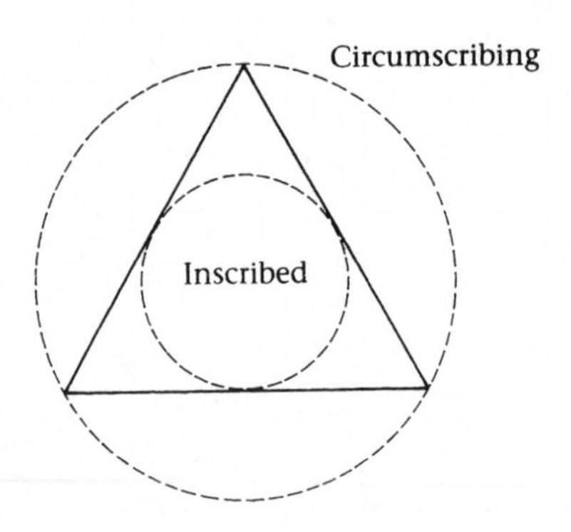

Max closed the book and moved to the fifth lectern, where he immediately announced, 'Book V is the most famous of the thirteen in the *Elements* and is known as the "Book of Proportions".'

Mr Ruche explained: 'What Euclid tried to do in Book V was establish the ratio between sizes: whether those sizes are geometric, in which case he was dealing with lines, area or volumes; or arithmetic, where he was dealing with numbers. Pythagoreans could not imagine a relationship between areas which were incommensurable. Euclid put an end to this notion, though he was simply developing an idea put forward previously by Eudoxus of Cnidus, a mathematician and astronomer.'

Standing in front of the sixth lectern, Max began again: '"The Book of Similarities".'

Mr Ruche continued, 'It is sometimes difficult to describe a shape, but we can easily tell when objects are the same shape.'

Max added, 'They might be the same shape, but they are not necessarily the same size.'

'Yes,' agreed Mr Ruche, 'they are the same in all but dimensions. So, when can two figures be said to be the same?'

Max, who was prepared, was waiting for the question, but it was Sidney who answered 'When one is proportional to the other.'

'And when can they be said to be proportional?'

Sidney replied, 'When their corresponding angles are...proportional and when the sides...the sides are...' Sidney was clearly confused.

Max interrupted: 'It's not his fault, it's just the text is a bit complicated. What Sidney was trying to say was, "When their corresponding sides are proportional and when their angles are equal."'

Seven books to go. Mr Ruche pressed on: 'Euclid's *Elements* draws on the work of mathematicians who went before – his second proposition is in fact Thales' theorem.'

Max moved to the seventh lectern and announced, Books VII, VIII and IX: the three books of arithmetic.'

Mr Ruche took up the story. 'In these books Euclid developed the work the Pythagoreans had done on whole numbers. We have already established that one of the primary functions of the mathematician is to catalogue things, firstly, into odd and even numbers. You remember your elegant definition, Lea – "those who believed in the number 2 and those who didn't". Even numbers can be divided into two equal parts, while odd numbers cannot. Then there are numbers which cannot be divided by any number. They are called *prime numbers*, so named because no other number can define them.'

Mr Ruche paused, remembering something that Grosrouvre had said in one of his letters: 'What is the sum of those things that define you or me? I think perhaps the time is coming when we will find out.' It took a moment or two before he came to again. Max, who had noticed what was happening, said, 'Second classification!'

Mr Ruche resumed. 'Second classification: divisible or prime. Prime numbers were to become one of the keystones of arithmetic. There are an infinite number of them. The thing that most surprised me about Euclid was that he didn't care about addition at all. What he cared about was division.

'A whole number can be arrived at only by the multiplication of its prime factors. Prime factors are the smallest prime numbers that divide into the whole number. Euclid's most famous work was to separate whole numbers into their prime factors. He went further than this and found the factors of two whole numbers a and b, and worked out their highest common factor – the largest number that can exactly divide both a and b – and their lowest common multiple – the smallest number that has both a and b as factors.'

Mr Ruche turned on the overhead projector. A strange drawing appeared on screen:

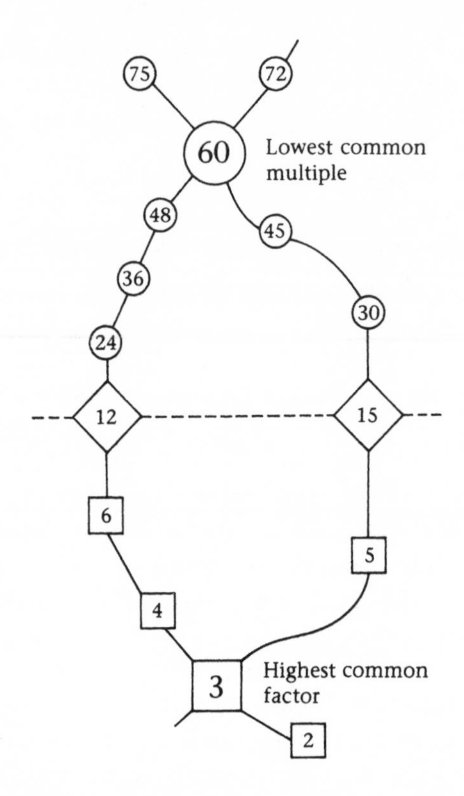

'Excellent, excellent!' Mr Ruche nodded in approval when he saw the drawing that Max had done earlier in the afternoon.

Max explained: 'If *a* equals 12 and *b* equals 15, then the largest prime number which divides into both – the highest common factor – is 3. And the smallest number that has 12 and 15 as factors – the lowest common multiple – is 60.'

Max turned off the lights over the three lecterns which had just been explained and moved on to the next: 'Book X, "The Book of Irrationals".'

Mr Ruche continued with his performance: 'In Book X, Euclid is using much of the work of Theodorus, the founder of the theory of the incommensurable. Theodorus dealt not only with straight lines which cannot be measured, but with those which can and with the areas of squares or rectangles of which these lines formed the boundaries. While the Pythagoreans knew of only one irrational number, the square root of 2, Theodorus had demonstrated that all whole numbers up to 17 had irrational roots – with the exceptions of 1, 4, 9 and 16, whose roots were known. Why he stopped at 17 we don't know. Theaetetus, however, picked up the thread and dealt with the numbers which came after. I have to say that this is easily the most difficult of the thirteen books, which is why it is known as the "Mathematician's Cross".

'Euclid is not concerned only with flat geometry. In Books XI and XII he goes on to identify the different forms in three-dimensional geometry: *solids*. Pyramids, prisms, cones, cylinders and of course spheres, as well as the regular polyhedrons. He calculated the surface areas and the ratios between their volumes.

'To do this, Euclid used a highly effective method developed by Eudoxus which later came to be known as the *method of exhaustion*. "Exhaustion" here means covering all the possibilities by a process of thought. To prove that two areas or volumes are equal, Euclid shows that the difference between them is smaller than any given area or volume. The process is not done step by step, it goes on endlessly and so "exhausts all possibilities" without having to define each one.

'To determine the area of a circle, for example, you draw a square inside it, then double the number of sides to make an octagon, which is closer to the size of the circle. Each time you double the number of sides of the polygon, its area becomes closer to the area of the circle, though the areas will never be exactly the same.' A slide appeared:

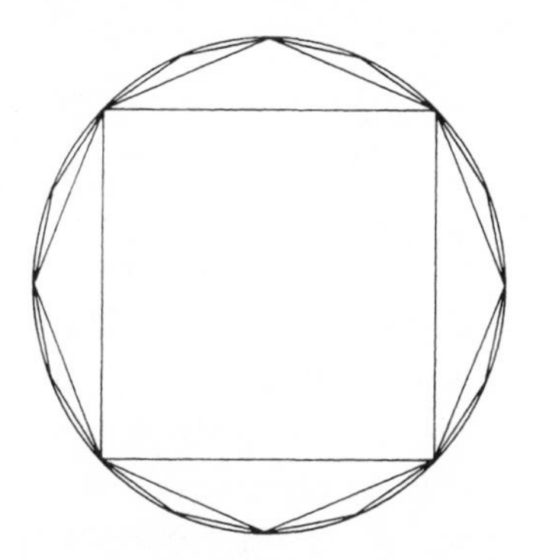

'The important thing is that the area of the polygon – which you can calculate – and the area of the circle – which you are trying to calculate – get closer and closer as you multiply the sides of the polygon. This means that you can find the area of the circle with greater and greater precision, though you can never find it exactly.'

Max moved along to the last lectern. 'Book XIII, the crowning glory of the *Elements*.'

Mr Ruche explained: 'In this book, Euclid explains what the previous twelve have been leading towards. He now has the tools to define the five regular polyhedrons which can be inscribed in a sphere. They are: the tetrahedron, a pyramid with a triangular base; the cube, with six square sides; the octahedron, in fact two equal pyramids joined at their square bases; the dodecahedron, with twelve facets, each a regular pentagon; and the icosahedron, with twenty facets, each an equilateral triangle.' Another slide appeared:

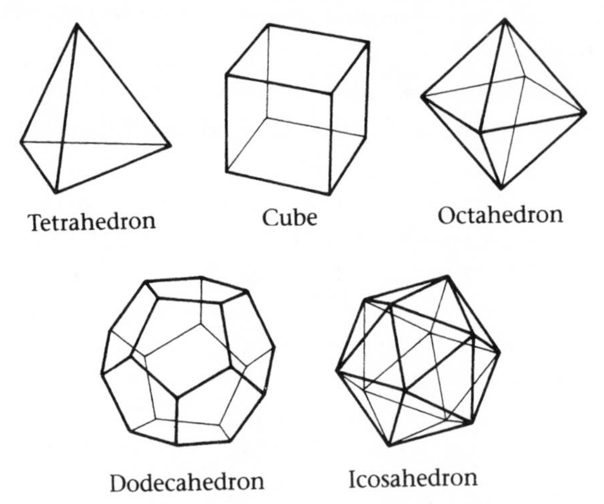

'Why five solids, why not four or six?' Jon-and-Lea chorused.

'Good question. That's the most extraordinary part: among the infinite number of polyhedrons, only five are regular. Usually, if one looks for a specific property in a series of mathematical objects, either there are none or one or an infinite number. For example, in the last diagram, an infinite number of regular polygons could be inscribed in the circle, but in three-dimensional space there are only five polyhedra. Needless to say, this puzzled the Greek thinkers. Plato had an answer for it: he thought that the five polyhedra represent the four elements in the universe – earth, air, fire and water – and the fifth, which was closest to the circle, represented the cosmos as a whole.* The perfection of each solid represented a specific element and the sphere in which they are inscribed, represented the universe, symbolizing its perfect harmony. These five polyhedra were known as the *Platonic solids*. Euclid's conclusion was this:

"There are no regular polyhedrons other than these five." '

Max moved away from the last of the lecterns and turned towards his audience. 'The thirteen books of Euclid's *Elements* comprised the basic knowledge without which no young man could become a mathematician in 300 BC.'

'Have you read the whole thing?' Lea asked Mr Ruche admiringly. 'All thirteen books?'

He would have liked to tell her that he had. At his age, it was so unusual to win the admiration of the younger generation that he couldn't help but seize it. 'Yes, yes I have', he lied. 'Well, that's the end of the *Elements*!' In fact, Mr Ruche intended to elaborate on the aim of Euclid's project. He pressed the button on the tape recorder and a voice came from the loudspeaker:

'Attention! Attention! No mathematical proposition should be accepted without proof; I repeat, without proof!'

'This was an unwritten rule that Greek mathematicians had imposed on themselves. But how did one go about proving a proposition? By deducing it from one which was already proved to be true.'

'But that's a vicious circle', said Max. 'Doesn't that mean that maths would just go round and round in circles?'

'That depends on where you start', said Mr Ruche. 'It's a very

*Plato takes the four elements of which all matter is composed – earth, water, air and fire – and reduces them to triangles. These fundamental triangles, if suitably combined, can be made into the only five geometrical solids known to exist. Plato then assigned each of these five solids to a different element: fire was associated with the tetrahedron (four equilateral triangles), the smallest, sharpest and most mobile of the regular solids; air with the octahedron (eight equilateral triangles); water with the icosahedron (twenty equilateral triangles); and earth with the cube (six squares), the most stable of the regular solids. The fifth solid, the dodecahedron (twelve pentagons), is closest in form to the sphere and is therefore identified with the cosmos as a whole.

delicate question.' As soon as he had said this, he knew the twins would be onto him immediately. He was right:

'I bet it is a delicate question, but you have to start somewhere', said Lea. 'Actually, it reminds me of something I read by someone called Polybius: "A good start is half the work." That means that if you get off to a bad start, you could be in for a tough time.'

'You're right!' Mr Ruche said, all fired up again. 'The only way of breaking the vicious circle is to have a number of statements which are accepted a priori and which cannot be later modified. We start with definitions. These come at the beginning and testify to the existence of the basic mathematical elements that will be used to create others, multiplying and filling the mathematical universe with new forms.'

'It all sounds a bit like the Bible, Mr Ruche. In the beginning... actually, before the beginning there was God and then God decides that Adam exists. He proposes a statement like "Adam exists; Adam is a man." Then Adam creates Eve out of one of his ribs or something. Then Adam and Eve create Cain and Abel and everyone was fruitful and multiplied.'

Mr Ruche listened, bemused, to Jonathan's version of the Bible – an axiomatic Bible. 'I've never really been religious, Jonathan.'

'Well, neither are we, but you have to read the classics.'

'The classics; does that mean you've read the Bible?'

'No, but then I bet you haven't read the *Elements* either...'

'Let's get back to Euclid', said Mr Ruche, quickly. 'After the definitions come the postulates and the axioms. The postulates state *a priori* that certain mathematical constructs can exist, while the axioms are commonly accepted ideas which form the basis of logical thought and which do not need to be proved. For example, if you had two objects, each of which was equal to a third object, it is impossible for them not to be equal to each other. An exact duplicate of an object is, by definition, the same shape as the original. This is why Euclid feels he can reasonably establish a whole list of axioms, many of which are relevant outside mathematics.'

Max turned on the overhead projector, and a slide appeared which listed Euclid's axioms:

Things which are equal to the same thing are also equal to one another.

If equals are added to equals, the wholes are equal.

If equals are subtracted from equals, the remainders are equal.

Things which coincide with one another are equal to one another.

The whole is greater than the part.

Mr Ruche expanded on this. 'Imagine two halves being different from each other. That would mean you had a whole that was lopsided. It's because they *are* equal that they are called halves. Those then, are the axioms. What are they used for? To make comparisons. Comparing halves to each other, parts to the whole, objects to one another, and so on. Without these, it would be impossible to make any comparison.

'On to the postulates.' Mr Ruche dropped his voice to a whisper. 'The thing that surprised me at first', he confided, 'was finding out that there are no postulates in arithmetic, only in geometry.'

'Because he didn't need postulates in arithmetic,' said Lea, 'otherwise I'm sure there would be plenty, like "between any two numbers there is always a third number" or "numbers are everywhere" or "if you add something to a number, the result is a number" or even "one number good, two numbers better, three numbers, all hell breaks loose!"'

Lea had to stop because everyone around her had collapsed into helpless laughter, as much from exhaustion as from what she had said. Slowly, they had moved away from the wonders of Alexandria, the lighthouse and the Museum, and on to the difficult theories of the time. They were tired and losing interest. The session had gone on too long.

'Euclid set out five postulates in geometry.'

'Like the five polyhedra?' asked Jonathan.

'Nothing to do with the polyhedra,' Mr Ruche replied

impatiently, 'or with the fact that, like most mathematicians, he had five fingers. The number is just coincidental. You know the first postulate.'

Click. The next slide appeared:

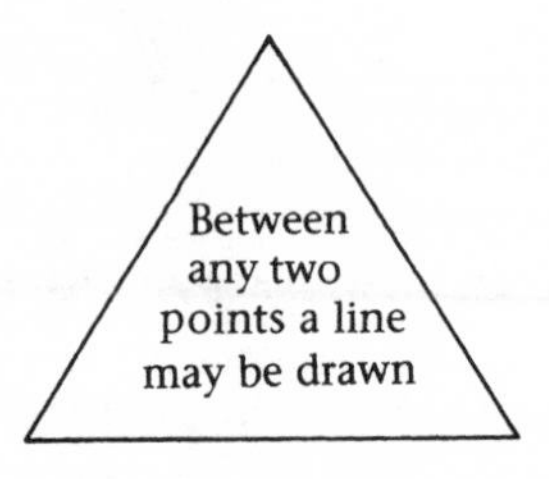

'What is Euclid trying to say in posing this? Regardless of where two points are in space, a line can be drawn between them without having to go around anything.' *Click*.

Any straight line may be continued indefinitely

'What is Euclid trying to describe here? That a finite straight line – a segment – can be extended as far as you like, as long as you have enough space. In fact, Euclid is trying to say that space is infinite in every direction. After lines, circles: the third postulate.' *Click*.

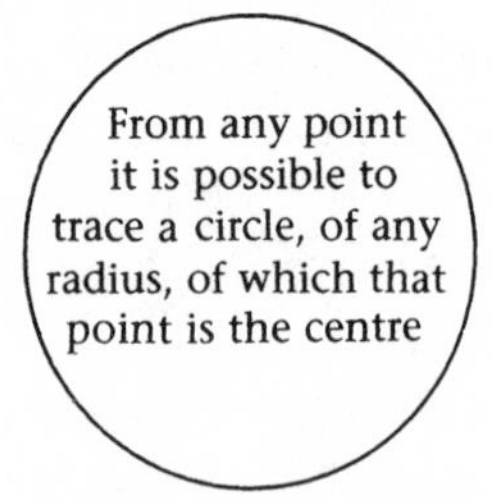

'By this, Euclid means that circles can exist anywhere and everywhere, and not only in some special part of space; and that circles can be as large or as small as you like. In the fourth postulate, Euclid moves on to angles.' *Click*.

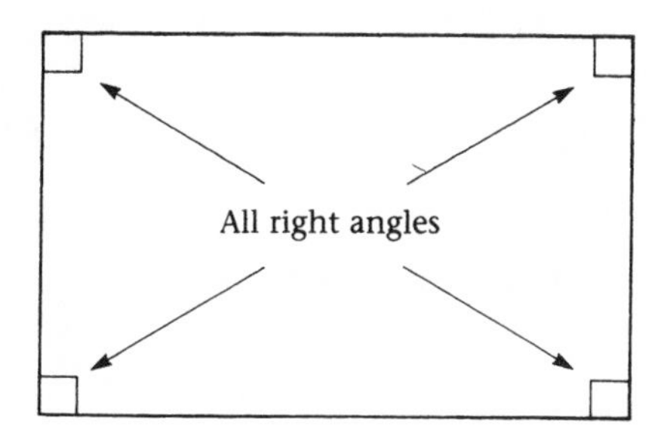

'Euclid wants to establish that right angles are always equal to one another, regardless of where they appear.'

'But what else could they be – they could hardly get bigger or smaller, could they?' Jonathan asked.

'That's exactly what Euclid is establishing – that they do not change at all.'

Max was out for the count, so Mr Ruche had to operate the projector himself: 'Lastly, there is the postulate of Parallel Lines which states' (*click*):

> *For a specific line on a surface and a point not on that line, only one line can pass through that point which is parallel to the first line.*

'And that says...well, what it means', added Mr Ruche.

'You mean, it means what it says', said Jonathan.

Everyone laughed. It was then that Perrette arrived. Needless to say, when Lea finally managed to explain that they were laughing at Euclid's fifth postulate, she looked at them as though they had all gone mad. All she could say was, 'And you think that's funny?' at which Mr Ruche, Max, Lea and Jonathan exploded with laughter again. Across the courtyard, in the dining room, Sidney heard the proceedings and began to squawk loudly. 'I wonder if parrots can laugh?' thought Perrette.

CHAPTER 10

One Risotto and a Chicken Korma

Apollonius and Hipparchus; trigonometry unleashed; a good argument & a significant curry

Next evening they met again in the screening room. Mr Ruche dimmed the lights, and suddenly a circle appeared on one wall. Max was holding a lamp perpendicular to the wall and the light, focused by a conical lampshade, fell in a perfect circle. From out of the darkness, Sidney's voice announced, 'Circle!'

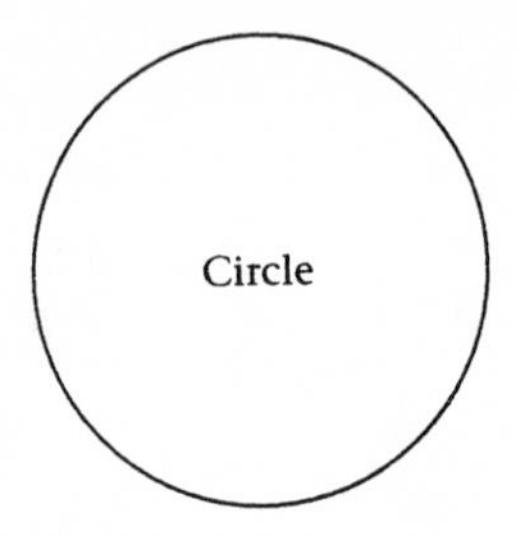

Max tilted the lamp to one side and the light now fell into an oval.

'Ellipse!' said Sidney.

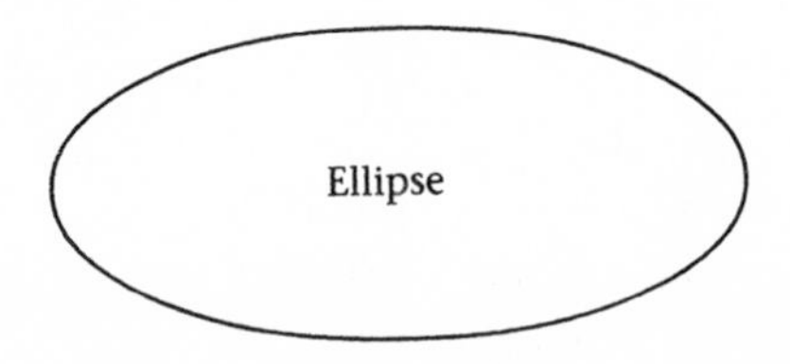

Max continued to turn the lamp and the shape grew and grew, before suddenly breaking, so that it no longer formed a complete ellipse, but stretched the length of the wall. 'Parabola!' said Sidney.

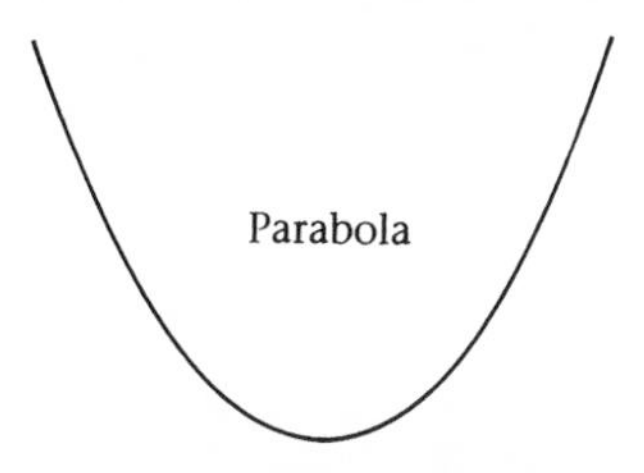

Now Max edged the lampshade away from the wall and the parabola grew until a second spot of light appeared on the far side of the wall. Sidney announced, 'Hyperbola!'

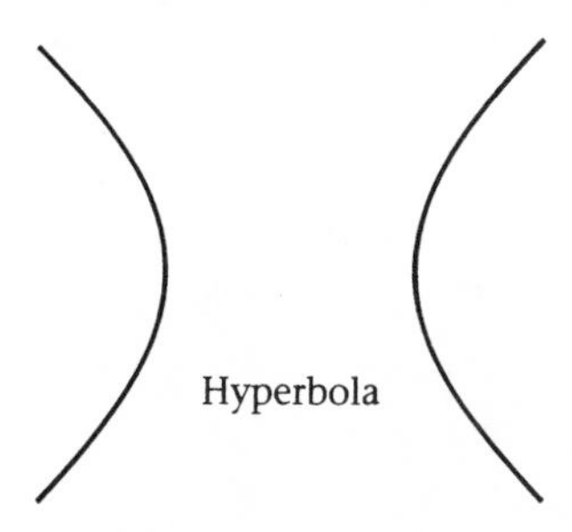

Mr Ruche stepped in at this point. 'You have just witnessed a meeting: the meeting of a cone of light from the lamp with the plane of the wall. This is why the four shapes you have just seen are called *conical sections*. They were discovered in the fifth century BC by Menaechmus, a Greek mathematician. Imagine his surprise at finding that all four figures could be produced by the meeting of a cone with a flat plane. You could move from one to another simply by rotating the cone on its axis!'

The twins' eyes lit up with surprise and pleasure, but also puzzlement. They could not equate what Mr Ruche had said with what they knew about a cone. He pressed the play button on the tape recorder and a voice intoned:

'Attention! Attention! This is a definition: a cone is a geometric figure defined by two lines, known as *generators*, passing through a point, the *apex*, on a *base* which is a circle.'

'Despite what you might think,' said Mr Ruche, 'a cone is made up of two parts, extending out on either side of the apex. What is commonly called a cone is in fact a semi-cone.'

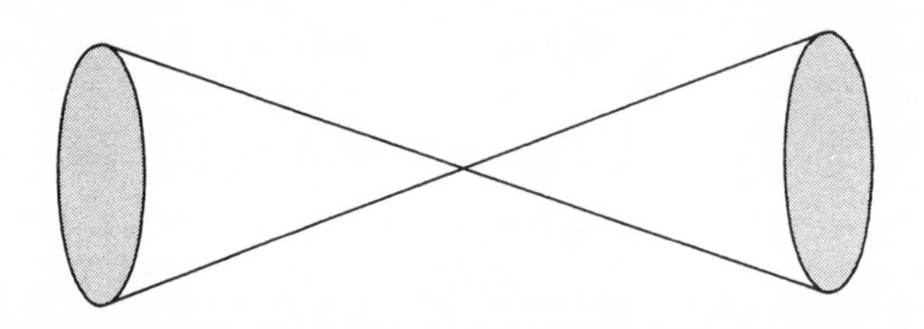

'You mean that every time I asked for an ice-cream cone as a kid, people were palming me off with half-cones?' Jonathan said indignantly.

'Just as well', retorted Lea. 'Try eating a real cone. While you were licking one end, the other end would go splat onto your sandals.'

Mr Ruche went back to his story. 'Two centuries later, Apollonius began work on the subject. He was called "the Great Geometer" by later mathematicians. He lived in Alexandria in the latter part of the third century BC and studied in the Great Library of Alexandria. He probably lived at the Museum. His great work, *Conical Sections*, consisted of eight volumes, only seven of which have ever been found. It was Apollonius who gave the conical sections their names. He created the words "hyperbola", from *hyper*, meaning "some added", "ellipse", from the Greek for "something missing", and "parabola", from the Greek word for "same", *paros*.

'These geometric curves appear everywhere in the natural world. Take the planets, for example. In the fixed sphere of the heavens they were the only bodies thought to move. Since the dawn of time, people have wanted to know about their movements.

'The Greek theory of harmony would have it that everything moved in perfect circles or spheres. Greek astronomers, particularly Eudoxus, even doctored their calculations to make it seem as though they did, but nature can't be tricked into doing what we want. The planets orbited the sun as they always had. Two thousand years after Eudoxus, Kepler discovered that the planets' orbits were elliptical rather than circular.

'The next great Greek mathematician is Hipparchus. I'll read you Grosrouvre's notes:

> Hipparchus is generally considered to be the father of trigonometry. Following on from the work of Babylonian astronomers, it was he who

divided the circle into 360 degrees. He was a noted astronomer and his long, detailed study of the heavens led him to establish the first "tables of chords", which were the most important tools for astronomers for centuries. The tables were so precise that by using them Hipparchus was able to work out that the earth's axis was not fixed but described a circle, returning to its starting point every 26,000 years. This was called the *precession of the equinoxes*.

N.B. If the earth's axis moves, then obviously the earth moves, so it is strange that Hipparchus maintained that the earth was fixed...and many others continued to believe it.

'What was happening in Alexandria? Ptolemy I, known as Soter, was succeeded by Ptolemy II. By the first century BC, Ptolemy IX was on the throne. There was no Ptolemy X, Ptolemy XI was murdered in an uprising, and Ptolemy XII was hounded from Alexandria and fled to Rome. When he returned, it was with Roman armies which became an occupying force and ended Egypt's independence.

'Ptolemy XII decided that his son, then aged ten, would succeed him as Ptolemy XIII, on condition that he marry his sister' – Mr Ruche paused for effect – his older sister, Cleopatra. They were married, but the marriage was not a happy one and Cleopatra fled Alexandria for Rome and returned with Julius Caesar. There was a revolt, and the populace of Alexandria laid siege to the lovers' palace.'

'What's he going on about all this for?' whispered Jonathan.

'Oh, I'm sure he's working up to something', murmured Lea.

'Caesar planned their escape. To make sure his fleet would not be caught, he had the ships in the great port burned. The fire spread to the harbour and from there to the Great Library of Alexandria. Tens of thousands of rolls of papyrus were burned, volumes which had cost the early librarians much time and effort to get.'

The twins glanced at each other – so this was what it was leading to.

'Caesar's ruse worked: Ptolemy's fleet was burned, but much of the library was burned with it' said Mr Ruche, sadly. 'A battle ensued between Caesar's troops and those loyal to the Pharaoh. Ptolemy XIII was killed, leaving Cleopatra a widow – though not for long. She had another brother, whom she married and who became Ptolemy XIV. He soon disappeared, probably killed on the orders of Cleopatra. When Caesar returned to Rome, Cleopatra followed, but she returned to Alexandria when Caesar was assassinated. It was not long before she fell in love with another Roman general.'

'Mark Anthony', chorused Jon-and-Lea. 'They were lovers and had three children.'

'I see that you know all about the Ptolemaic dynasties!' said Mr Ruche, surprised.

'You're not so bad yourself – it's just that we've seen every film about Cleopatra ever.'

Mr Ruche resumed his story. 'Cleopatra was determined to rebuild the Great Library of Alexandria. Mark Antony pillaged the library at Pergamum and brought 200,000 volumes to Egypt as a gift to Cleopatra, where they joined those which had been saved from the fire. Cleopatra was the last queen of Egypt. After Egypt was conquered in the Battle of Actium, Cleopatra committed suicide and the country became a province of the Roman Empire.

'Alexandria, however, continued to be a refuge for scientists and scholars. Two, in particular, are important for the influence of their work: Ptolemy in the second century AD and Diophantus in the third. Claudius Ptolemy – no relation of the kings of Egypt – is better known as an astronomer than as a mathematician, although his work deals more with maths than with astronomy. His great work was the *Mathematical Syntax*.'

Max held up Grosrouvre's copy and announced, 'Mathematical Syntax: thirteen volumes.'

The twins looked panicked for a moment, worried that Max was going to go through each volume one by one, just as he had for Euclid, but he simply read out the index card with Grosrouvre's notes:

> At this time, astronomy was the science of 'explaining the universe'. It attempted to chart the movement of heavenly bodies and to give them a geometric explanation. Eudoxus, Hipparchus and Ptolemy all tried to construct a mathematical model of the universe to explain the movements of the stars and planets.

Ptolemy places the earth at the centre of his system. It does not move, but everything else moves round it. The heavens, to him, were lit up by circles and spheres, and his treatise was developed around the geometry of the circle and the sphere.

Mr Ruche thumbed through his notebook and resumed his history: 'The Roman Empire crumbled and the Byzantine Empire took its place. Alexandria, previously a pagan enclave, became Christian. Science, which was much prized in Greece, was neglected by Rome. The only art which counted on the banks of the Tiber was the art of government; the Romans were passionate about laws, but not about the laws of mathematics. In a thousand years of the Roman Empire, there is not a whisper of mathematical thought.

'The combination of the Roman lack of interest in pure thought and the Christian rejection of ideas which were not directly linked to religion had tragic consequences for the survival of the sciences. The first to suffer was Hypatia, the first great female mathematician.'

At this, Lea, who was getting bored, pricked up her ears.

'At the end of the fourth century, a family of famous mathematicians lived in Alexandria: Theon and his two children Hypatia and Epiphanus. It was Theon who first set down the method of calculation of a square root, while Hypatia, his daughter, did remarkable work extrapolating the discoveries of Apollonius, Diophantus and Ptolemy. Epiphanus also worked on Ptolemy's astronomical theories, but was less talented than his sister.

'Like her Greek colleagues of old, Hypatia was not only a mathematician but a philosopher and taught both subjects. Hundreds of pupils flocked to her lessons, in awe of her intelligence, her knowledge and her beauty. All of these were anathema to the new moral order which ruled Alexandria. Hypatia was a liberated woman.

'In AD 415 the people of Alexandria attacked her chariot, stripped Hypatia and carried her off. She was tortured with oyster shells sharpened like knives and burned alive. She was not the only woman to be burned at the stake; religious authority often felt that a woman was better dead than powerful.'

Lea looked at Mr Ruche, pale. 'One female mathematician in all of the ancient world, and she's tortured and burned.' Lea was furious. 'And people wonder why girls don't do maths!'

Mr Ruche went on, charting the ineluctable decline of the ancient world: 'So, Greek mathematics was finished. Rome had only one mathematician, Boethius. Later, the Emperor Justinian ordered that all the "pagan universities" – including the Academy and all the Athenian schools – be closed.

'In 642 Arab forces seized Alexandria and it became a Muslim city. Three years later, riots in the city resulted in the burning of thousands of the Library's books. This was the end of the Great Library of Alexandria. The Museum survived for some time, but in 718 Omar II ordered the scientists working there to move to Antioch. That was the end of Alexandria.'

The session was at an end, too.

The following day, Lea went to see Mr Ruche. She arrived just as he was about to set off for the market at Abbesses. She was still upset about the death of Hypatia, about the destruction of everything: the Library, the Museum and the ancient world itself. Without realizing it, she had become attached to this world in the weeks since she had started to learn about it. Now it was gone. Questions teemed in her head, the first and most important of which she asked Mr Ruche: 'Why did maths begin in Greece and not somewhere else?'

He considered the question again as he wandered around the market buying veal and vegetables to make osso bucco – it was his favourite dish. Indeed, this was a question he had asked himself many times. As he wheeled himself back towards the bookshop he came up with an answer which he thought explained it: the Greeks liked to argue.

Osso bucco is served with saffron risotto and gremolata. He started with a saucepan of chicken stock. The stock is the most important part of the recipe, and Mr Ruche was careful to get it just right. When the stock began to boil he ladled a little into a bowl and plunged the strands of saffron into it, then turned down the heat so that the stock would continue to simmer gently. It was pouring with rain outside. The rain drummed steadily on the kitchen window as Mr Ruche prepared the knuckles of veal. Lea listened as he gave her the answer which had come to him on his way back from the market. He put a copper pan on the heat, cut three generous knobs of butter and added it to the pan. When it started to sputter, he put in the knuckles of veal. As he was adding the last, Lea interrupted him:

'So you're saying the Greeks discovered maths because they liked to argue. For that last ten years, the only thing I've heard from my maths teacher is "Miss Liard, I will have no arguing in my class." '

Mr Ruche felt he owed her an explanation. 'Yes, the Greeks liked to argue, but it's not the same as fighting. In ancient Greece, arguing was a noble pastime which allowed you to convince someone of your point of view.'

When the veal had browned, Mr Ruche lifted it onto a large plate. He pointed to the net of onions on the shelf above. Lea took out two and sliced them thinly.

'In the stadium, athletes challenged one another in the games, in the public squares they challenged each other with words. Both contests were governed by strict rules.'

Mr Ruche added the onions to the pan with celery, carrots and a little stock from the saucepan. He put the veal back in the pan, added parsley and some tomatoes, and left it to cook. As Lea dried her hands, Mr Ruche talked about Greece and the art of argument. He described the men in the ports, sitting at small wooden tables, talking, all of them listening and each in turn making his point, sipping a glass of ouzo, eating calamari and sliced scarlet tomatoes.

'I didn't know if they had ouzo in Thales' and Pythagoras' day, but I'm sure they ate grilled calamari and the conversation wouldn't have been so very different.'

The osso bucco was cooking nicely. It was time to start preparing the risotto. Lea took down a bottle of oil and handed it to Mr Ruche. It was the finest, cold-pressed, extra virgin olive oil, straight from Tuscany. Lea wiped her hands as Mr Ruche poured a cup of the oil, holding the slippery bottle carefully.

'Greek argument was never vague. It was a system which worked in stages. The more I think about it, the more invaluable I believe it is.' He pointed a finger at Lea. 'Do you accept that all men are mortal?'

Lea was surprised at first, but she was intrigued. 'Yes,' she said solemnly, 'I accept it.'

'And do you accept that Socrates is a man?'

'Yes,' she said, 'I accept it.'

Mr Ruche clapped his hands: 'There, that's all there is to it. All men are mortal; Socrates is a man; therefore Socrates is mortal. And there's nothing you can do about it, Lea. You've accepted the first two, so you must accept the third!'

The stock was bubbling. Lea took the heavy pan which hung nearby and put it on the stove. Mr Ruche chopped two shallots, poured a little oil into the pan and turned down the heat.

'It's not that what you're saying isn't interesting, but with all this talk of Greece and calamari and Socrates, you haven't answered my question: why did maths originate in Greece?' asked Lea.

'I'm getting there, I'm getting there. Thales, Pythagoras, Hippasus of Metapontum, Hippocrates of Chios, Democritus, Theaetetus, Archytas of Tarentum – all the Greek thinkers who worked in maths – who were they, what did they do, where did they figure in society? They didn't work for the state like the mathematicians of Babylon and Egypt, who merely calculated and belonged to the same caste as the priests and the scribes. For the Greek thinkers there was no high priest to limit their thoughts and ideas. Greek thinkers were free men. But...' The shallots were softening in the heavy pan. '...but they had to be able to defend their point of view against their peers.'

Mr Ruche explained to Lea that each thinker was recognized as an indivdual, even those who belonged to a 'school'. Such emphasis on the individual was almost unheard of previously. They used this personal freedom to think and to advance their theses and develop their theories. They were accountable for their work, not to a higher authority but to anyone who might contest, criticize or contradict it. Though in everyday life they were constrained by the law like any other citizen, in the world of ideas they were free men.

'Greece at the time was a constellation of independent city states. Some were tyrannical, others democratic. In the latter, the citizens actively participated in political life. In Athens, the meetings of the assembly sometimes had seven or eight thousand people, and each one of them had the right to speak and try to convince others of their point of view. Can you imagine? At the end of a meeting they would vote. Even Greek courts did not appeal to the law of God, or of a king, but to a judge and jury made up of the people, who had to be convinced.'

'Yes, but what about maths – you're getting off the point.'

'I'm not off the point at all. This is the point.'

Mr Ruche lifted the lids of the pans in turn: the veal was cooking in the frying pan and the shallots softening in the saucepan.

'You can argue successfully only if you are agreed on the basis for the argument. Once that is agreed, you can discuss the rest. I say something, you respond, I make a point, you argue the point, you refine your argument, I shift my ground. But in the end, which one of us is right? How do we settle things, who has the last word?

'Greek philosophers, politicians and lawyers were skilled in the art of persuasion, but persuasion cannot eliminate all doubt. Mathematics required something more than just persuasion – its arguments had to be irrefutable. Mathematicians had to be able to make statements that no one would be able to disprove. They wanted absolute proof. It was this that marked out Greek thinkers from mathematicians at the time.

'Unlike the Egyptians and the Babylonians, the Greeks refused to accept that intuition was enough to support mathematical truths. They also refused to accept specific examples as proof: I believe because I see, you believe because I show you. This was the "proof" accepted on the banks of the Nile and the Euphrates. Greek mathematicians were not satisfied with this: they wanted an argument, a proof.'

'Didn't that exist before them?' Lea asked, surprised.

'No, the Greeks invented it.' Mr Ruche added the rice to the shallots and stirred until the grains became translucent. He kept stirring, careful not to let the grains stick together. When he had a rhythm going, he went on: 'But turning your back on intuition and on empirical evidence means there is always a doubt. If seeing is not believing, then how can I ever state that something is true? How can I convince you, or even myself, that what I say is true? This raises the most important questions the Greeks asked themselves: How do we think? Why do we think what we think? How do I prove that what I think is true?'

From his tone, Lea could tell that they were important to Mr Ruche, too. They were questions that had never even occurred to her.

'In order to deal with the doubts that this raised, Greek thinkers developed systems to counter the doubts', Mr Ruche went on, trying to divide his attention between the conversation and the delicate preparation of the osso bucco. 'These systems were intended to justify the assertions they made. For the first time in human history, man thought about the process of thought itself.

'Aristotle developed these ideas in a book he called *Organon* – 'The Tool' – by which he meant "logic". This was the birth of logic, which he considered should be the rules of thought. Logic would decide how to establish truth. The rules of logic were applied to each proposition. Since the process did not change and was independent of the subjects it dealt with, it was accepted as impartial.'

Mr Ruche ladled some stock into the heavy pan.

'The process was based on a number of simple principles, of which two are pre-eminent. The first is that "A statement and its opposite cannot both be affirmed." In other words, a proposition and its contradiction cannot both be true. This is the *principle of non contradiction*. The other principle is that "An assertion and its opposite cannot both be false." If one is false, the other must be true, there is no other possibility. This is the *principle of no third*. There!' concluded Mr Ruche. 'That's how the Greeks went from *showing* to *proving*.'

For the first time since he had started cooking, Mr Ruche looked at the recipe to check that he had done everything correctly.

'One more thing', he added. 'The development of the alphabet in Greece favoured the idea of written proofs. It is much easier to check that one has not made mistakes if the proof is written down, especially if it is long.'

Mr Ruche lifted the lids of the pans in turn: everything was perfect. By the time he rolled his wheelchair into the dining room with a tray on his lap, everyone was waiting. As he served the osso bucco and risotto, Lea went out to get the bottle of chianti which had been chilling on the balcony. They drank a toast to the Greek thinkers. '*Bon appétit*!' said Mr Ruche, taking a mouthful. They all enjoyed the meal, and it was very late before the light in the dining room went out.

To thank him for the supper, Jon-and-Lea invited Mr Ruche to the

Shalimar, the best Indian restaurant in the area. He was happy to accept. He wondered why they had chosen this restaurant, but he knew it would not be long before he found out. As the waiter brought their cocktails, Lea offered Mr Ruche his first clue and began to tell the story of an Indian mathematician.

'Līlāvatī had everything going for her. She was beautiful and intelligent. Her father was a famous astronomer. When she came of age to marry, he spent a long time consulting the stars. Her horoscope shocked him: if she married, he would die. Bhāskara, her father, loved life and so he forbade his daughter from finding a husband or leaving his side. But he did give her name to the book that contained his life's work: *Līlāvatī*. In it were a host of mathematical problems that he was the first to solve. Written in the twelfth century, *Līlāvatī* became one of the most famous of all Indian works of mathematics.'

'There is another version of the story', said Jonathan. 'In it, her horoscope suggested that Līlāvatī would marry, but her married life would be cut short. Bhāskara consulted the stars and decided her fate could be avoided if Līlāvatī were to marry on a certain day. To count off the days to her wedding, Bhāskara created an hourglass and Līlāvatī would often come and watch the sand trickle through. One day, as she leaned over, a tiny pearl fell from her nose ring into the sand, slowing the rate at which it trickled through. The wedding was celebrated some days after the appointed day. Soon afterwards, her husband was killed in an accident. To console her, her father dedicated his book to her.'

'Typical!' Lea's voice could be heard across the restaurant. 'It's all the woman's fault. Her pearl which slows down the hourglass and causes her husband's death. Lucky she had another man around to write a book and dedicate it to her. Trust you to come up with the macho version of the myth!'

'Yeah, well, men are to blame for everything in your book.'

'I suppose you brought me here to watch the two of you fight over different versions of the same myth', protested Mr Ruche.

'No, no,' said Jonathan. 'Tell him, Lea.'

'Remember you said everything started with Thales, that the Greeks invented maths? Well, what about the Babylonians, the Indians and the Chinese? Why isn't there a copy of Līlāvatī in the

Rainforest Library? And there are no transcriptions of Babylonian tablets, no Chinese or Mayan texts.'

Jonathan, meanwhile, leaned down and pulled a package from under the table. From it, he took a large book.

'Look at this', he said, handing the book to Mr Ruche. 'Ahmes, he was writing a thousand years before Thales. This version is a facsimile of the Rhind Papyrus, found in the tomb of Rameses II at Thebes. The papyrus is five yards long, and on it Ahmes solved a dozen mathematical problems. It is the oldest mathematical document ever found.

'Ahmes tells us that he was a scribe and that the text was written in the fourth month of the rainy season, in the thirty-third year of the reign of Apophis in the fifteenth Dynasty. Apparently that dates it to the sixteenth century BC, but he also says it is copied from something written in the reign of Ammenemes III, more than 2000 years BC. Some people think the mathematics in the Rhind Papyrus dates from the building of the pyramids back in 2800 BC.'

Not wanting to upset Mr Ruche too much, Lea suggested, 'Perhaps we could just agree that not everything began with Thales.' Mr Ruche smiled and nodded.

'Two thousand years BC there were other mathematicians', said Jonathan as he started on his tandoori chicken. 'In Babylonia, Mesopotamia and in Egypt. In China they were already writing proofs – not in the Greek style – but methods of demonstrating things they believed to be true about numbers and shapes.'

Lea nodded towards the book. 'When you read the Rhind Papyrus, you'll find out that Ahmes sets out to "explain the rules for studying nature and to learn everything that exists, every mystery and every secret."'

'Everything that exists!' Mr Ruche exclaimed. 'There's a lot of "everything" about!'

Lea had spent two nights poring over the hieroglyphics in Ahmes' book. 'The first six problems you see here', she said, pointing to a column of hieroglyphics, 'are about how to divide a certain number of loaves of bread – between one and nine – between ten men. Really, it was just a way of writing down the multiplication tables for the numbers 1 to 9.'

Mr Ruche felt a little emotional: this was the first present the twins had ever given him. Lea, however, was not to be deterred from her mission. 'You see here?' she said, turning the pages. 'In problem 50, Ahmes talks about squaring the circle and he calculates π as 3.16. That was 2000 years BC and he's only half a per cent out!'

'I'm sure you know,' Lea continued, 'that according to the Greek historian Herodotus, the true beginnings of geometry were in Egypt. In 2000 BC, Rameses II decided to give each of his subjects a plot of land of equal size so that they would each have to pay the same taxes. However, every year the Nile burst its banks and some plots would lose ground to the river. Rameses sent his scribes to measure the area lost, so that they could calculate the percentage of tax to rebate. This was the true beginning of geometry.'

'Thanks for reminding me. If I remember rightly, Herodotus said, "It is when we do not have equality that we need geometry."'

'To re-establish equality is to establish liberty', said Lea. 'You told me that the Greeks were free thinkers, Mr Ruche, and I think that that's the real difference. Everywhere else, mathematics developed in strict hierarchial societies.'

As they ate, they talked about Mesopotamia and Egypt, China, the Aztecs and the Mayans. When he got back to the Rainforest Library, Mr Ruche created a new section. Section 5: Mathematics Outside the Western World. And in it he placed the Rhind Papyrus.

CHAPTER 11

Problems, Problems, Problems

A plague in Athens; Hippias, Hippocrates and Plato & the three great mathematical problems of the ancient world

The investigation into Grosrouvre's death had not progressed one iota, as the Greeks would have put it. It was almost Christmas, and though Mr Ruche, the children – even Albert – had thought long and hard, nothing about the mystery of Grosrouvre's disappearance had been resolved. They still didn't know who his 'loyal friend' was, nor the identity of the gang who had tried to get hold of his proofs. Most importantly, they still didn't know whether Grosrouvre's death was accident, murder or suicide.

Mr Ruche had planned a new session to take place at their Christmas dinner, or rather just before it, during which he would tell them what he had learned from the Rainforest Library. Afterwards, they would discuss what they should do next.

Mr Ruche had decorated the dining room beautifully. Garlands of gold and silver stars were hung around the room from invisible threads. Perrette was determined not to miss this session, the last of the year; the twins would be leaving to go skiing the following day. They set a place for Albert, but he had warned them that he might not make it before dinner. 'It's not that I'm not keen,' he said, 'but Christmas Eve is a gold mine for a taxi driver.'

They started without him. Sidney opened the session in a grave voice: 'The three great problems of the ancient world were the squaring of the circle, duplicating the cube and trisecting an angle.'

Max, in a clear, calm voice, explained the three problems. 'Squaring the circle involves creating a square with the same area as a given circle. Duplicating the cube involves creating a second cube precisely twice the volume of the first. Trisecting an angle involves splitting the angle into three equal angles. The first concerns areas, the second, volumes and the third, angles.'

'Squaring the circle,' Sidney announced.

As Max installed himself behind the projector, Mr Ruche picked up the story. 'In Babylon and Egypt there was already great interest in the connection between the circle and the square. In the oldest mathematical text yet discovered,' he held up his copy of the Rhind Papyrus, 'Ahmes, a scribe, tried to determine "the square equal to a given circle". He proposed a square with each side equal to $\frac{8}{9}$ of the circle's diameter. It was a little approximate.

'Two thousand years later, in Greece, Anaxagoras was the first of the Greeks to interest himself in this question. Anaxagoras was a political prisoner when he set out to resolve the problem of squaring the circle. While his fellow prisoners jeered at him, he scribbled calculations on the wall of his cell. He didn't solve the problem.

'Thanks to the intervention of Pericles, the founder of Greek democracy, he was freed, but, unable to live with the humiliation, he committed suicide. The problem of squaring the circle survived him. Next to the task came Hippocrates of Chios.'

'The one obsessed with crescents of the moon?' asked Jonathan.

'Exactly!' said Mr Ruche. 'I thought you might remember him.'

'We hang onto your every word', said Jonathan.

'I mean', said Lea, not to be outdone, 'what do you think we are...?'

She stopped, embarrassed, as she caught Max's eye and shrugged in apology. Max nodded to let her know she could finish her sentence.

'...deaf?' she finished, in a small voice.

'Jonathan mentioned Hippocrates' obsession with lunes, and that is exactly what we are going to talk about. The fact that Hippocrates succeeded in squaring the crescent, or lune, was an enormous leap forward. Before him, the only shapes that had been squared were rectilinear: rectangles, parallelograms, trapezoids. In squaring a curved shape, Hippocrates raised the hopes of other mathematicians.

'Hippocrates tried and failed to square the circle, and the many Greek thinkers who followed failed too, but, as we shall see, the problem continued to interest mathematicians long after.'

Sidney beat his wings to get their attention, and announced: 'Doubling a cube!'

Mr Ruche went on, 'The first mention of duplicating the cube comes from a time when a plague was ravaging Athens. Nothing seemed able to stop it. A delegation of Athenians went to Delphi to ask the oracle how they might stop the epidemic. They waited, fearful that the gods, through the oracle, might not answer them. This was their last hope. Without divine intervention, Athens would be destroyed. At last, the oracle returned.'

Sidney beat his wings again, standing tall on his perch: 'Athenians! If you wish to stop the plague, you must double in size the altar dedicated to Apollo on the island of Delos.'

'The altar of Apollo was famous throughout Greece for many reasons, not least for its shape: it was a perfect cube', explained Mr Ruche.

'Doubling the size of the altar seemed simple to the Athenians', said Max from behind the projector. 'They sailed out to the island and built a new altar, the sides of which were twice the length of those of the first.'

'But the plague continued', said Mr Ruche, 'and the Athenians despaired, when a wise man passing by explained to them that the new altar was not double the original, but eight times bigger!'

Perrette looked confused as two cubes appeared, one enormous, the other tiny in comparison. Suddenly, she understood: 'Of course, 2 cubed is 8. I never thought of it like that. 2 squared means the area of a square with sides that are 2 units long, and 2 cubed is the volume of a cube with sides 2 units long.'

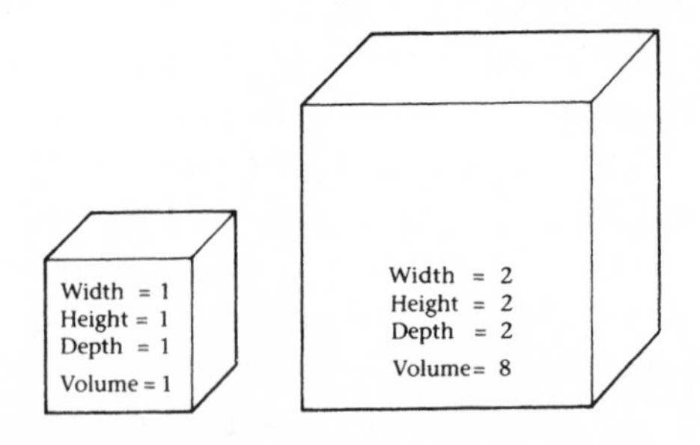

Jonathan looked at his mother, his eyes wide. It hadn't occurred to him that she could get excited about something as ordinary as a cube.

'Back on Delos,' continued Mr Ruche, 'the Athenians quickly constructed the new altar and set about doing exactly as the oracle had asked. On top of the original altar, they built another exactly alike.'

A slide appeared:

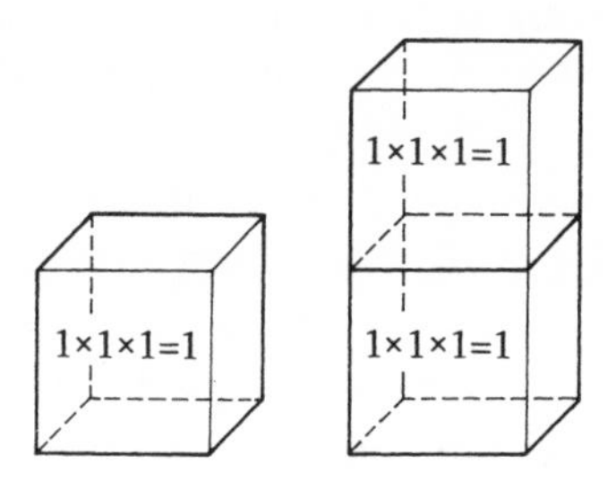

'The sum of the two altars is exactly twice that of the original altar', declared Mr Ruche. 'Satisfied, the Athenians returned home, but the plague continued. They were angry and confused: had they not built an altar which was a double of the original?'

'No. they hadn't!' Perrette was flushed with excitement, 'what they had done was not double the volume of the altar, but create two altars.'

Mr Ruche nodded and smiled. 'Exactly. The Athenians couldn't understand why they were unable to solve a problem that seemed so simple to them. They knew how to double a square. The educated among them knew that the easiest way to do this was to follow the diagonal.'

Max changed the slide:

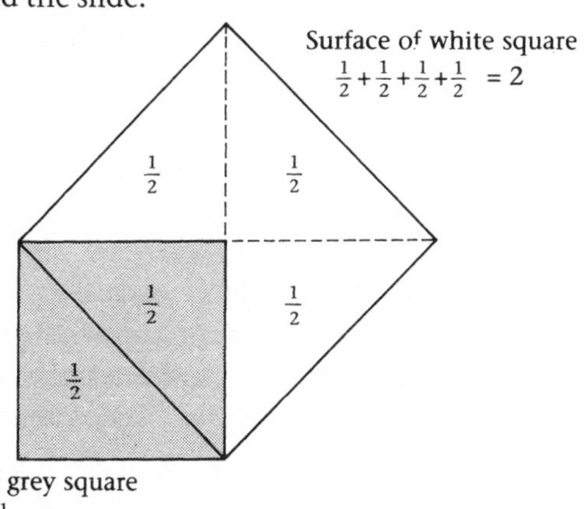

'But despite their best efforts, they could not double a cube', said Mr Ruche.

Perrette stood up. 'What about the plague? Had it stopped?'

Ignoring the question, Sidney announced, 'Trisecting an angle!'

Mr Ruche explained: 'The Athenians already knew how to divide an angle into two equal angles. The *bisector* had been discovered and was easy to use.'

Max knew this himself, having used it often enough in school.

'Dividing an angle into three couldn't really be much more complicated, they thought,' said Mr Ruche, 'especially as Thales' theorem had already shown them how to divide a line segment into three equal parts. But they were wrong. This problem also foxed Greek thinkers.'

'Mr Ruche, are you really saying the Greeks never managed to solve any of these problems?' asked Perrette.

'Not one', said Mr Ruche. 'Oh, there were Greek mathematicians who offered solutions: Hippias of Elia, Archytas, Eudoxus, Menaechmus, but only by breaking a fundamental law of Greek maths.'

'A law? What law?' Jonathan exploded. 'You never said anything about laws!' Jonathan had problems with authority, and the very mention of the word 'law' made him see bars in front of the windows.

'The Greeks were very concerned about tools. Mental tools, like logic, but also physical tools. The Greeks had refined their tools to include only the compass and the ruler. A ruler makes a straight line and a compass, a circle. You could not find anything more elemental, and they were interested in *elements*. With a single movement you could draw a line or a circle.'

Perrette looked at her watch. She was getting a little worried about Christmas dinner. The session would have to end soon.

'Therefore,' Mr Ruche announced solemnly, 'the three problems of the ancient world would be more accurately described thus: "*Using only a compass and a ruler,* construct a square equal in area to a given circle; construct a cube double the volume of a given cube; trisect an angle into three equal angles." Those words at the beginning change everything. As I said, some Greek mathematicians offered solutions to the problems, but they did not use just a compass and a ruler!'

Perrette wasn't listening, she seemed preoccupied. Suddenly she again asked, 'But what about the plague, Mr Ruche? You tell us this story about the cube, but actually it's a story about a terrible plague.'

'I haven't forgotten, I'm coming to that', said Mr Ruche. 'After they failed a second time they didn't know what to do. They were powerless. Desperate, they decided to ask the best mathematicians of the age. Among them, as I said, were some who offered solutions to the problem.

'Archytas approached the problem using the intersection of three surfaces: a cone, a cylinder and a torus. Menaechmus tried to resolve it using conic sections – a hyperbola and a parabola. But it was Hippias, the Sophist, who dared to break the rule of using compass and ruler.

'To Hippias all problems were technical problems. He didn't worry too much about theory, he wasn't afraid to use whatever method he needed and would try any trick he could think of to get what he wanted. It was this determination that helped him solve the most difficult of problems, by approaching them from a technical viewpoint. He managed to square the circle using the *trisectrix*, which he invented.' A slide appeared:

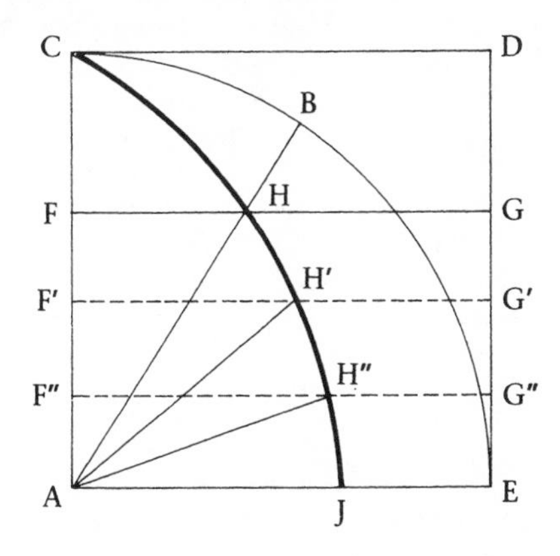

'Three centuries later, following his example, Diocles invented the *cissoid*.' Another slide appeared:

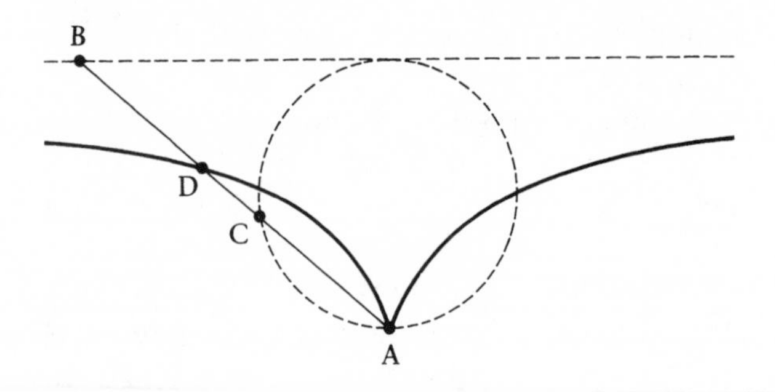

'This allowed him to solve the problem of trisecting an angle, and a century later Nicomedes invented a shell-shape which he called the *conchoid**, which was invaluable for duplicating the cube or trisecting an angle. Yet another slide appeared:

*The trisectrix, cissoid and conchoid are all curves which are generated by allowing a line AB to rotate about the point A. A point on AB whose position is determined in relation to another line then traces out the curve. For the quadratrix, AB is the radius of a circle. As it rotates at a steady speed from AC to AE, a line GH, parallel to CD, moves at a steady speed from CD to AE. Both AB and FG start and finish their moves at the same time. The trisectrix CJ is traced out by the point H, which is where AB and FG cross. It is easy to see how to trisect an angle, say the angle BAE. Look at the line FG: when it is one-third of the way towards AE, at position F'G', the line AH has moved to AH', one-third of the way round from AB to AE. And at position F"G", two-thirds of the way down, the radius line is at AH", two-thirds of the way round. Nearly a century after Hippias, Deinostratus showed how the same figure can be used to square the circle (though this is rather more complicated to illustrate). For this reason, the curve is also known as the *quadratrix*.

For the cissoid (the name means 'like ivy'), the line AB runs from A, on the circumference of a circle, to B, on a line which is tangential to the circle at a point diametrically opposite A. AB cuts the circle at C. As AB rotates about A, the cissoid is traced out by the point D, which moves in such a way that AD always equals CB. For the conchoid, B lies on a line which doesn't pass through A. As AB rotates about A, the conchoid is traced out by the two points C and D, which lie on AB extended in both directions and are equidistant from B: BC = BD. There are two parts ('branches') to the cissoid, one on either side of the straight line. How these two curves are used to trisect an angle is left as a challenge to the reader.

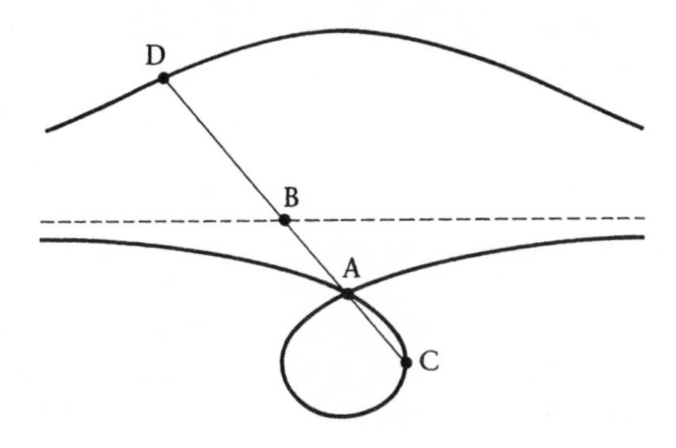

'But what about that plague? You keep forgetting about the plague!'

'Don't worry Perrette, we're getting there, I promise. All the curves invented by mathematicians to solve these problems are what are called *mechanical curves*. According to the laws of geometry at the time, these were inferior. They were outside the rules: they used the ideas of movement and speed. Points that move and lines that slide, shapes that shift – these were all prohibited. The world of Greek geometry was resolutely static.

'These shapes had another drawback – because they used movement, they could not be constructed,* and this posed a problem for the temple of Delos, since the Athenians had been commanded to build an altar. The mathematicians who invented these curves had not helped; the plague continued to ravage the city. The Athenians decided to consult a philosopher and went to speak to Plato at the Academy. This is what he told them: "If Apollo has told you, through the oracle, to build this new altar, it cannot be because he requires a bigger altar. He has asked this to reproach the Greeks for their neglect of mathematics and their disdain for geometry. In your haste to resolve this problem, you have been quick to abandon reason and have contented yourself with clumsy

*It is theoretically possible to construct a countable set of points on the curve using only a compass and a straight edge. However, this is not good enough to be a true construction. Although points on the quadratrix may be found as close to each other as desired, a continuous line cannot be generated in this way.

empirical solutions. In doing this, are you not forever losing that which is best in geometry?"'

Perrette opened her mouth to say something, but Mr Ruche quickly interrupted her: 'And at that moment, the plague ended.'

Time was up. It was almost time for Christmas dinner and there were still so many things to prepare.

When it came to Christmas dinner, Perrette never broke with tradition: she served foie gras, stuffed goose, mandarin oranges and Christmas cake. Mr Ruche had chosen the wines. A soft white claret with the foie gras, a solid red burgundy to go with the goose and, for the Christmas cake, an Épernay champagne.

Just as they were starting, the door opened. It was Albert. He was unrecognizable. Gone were the grey jacket and the cap. His hair was slicked back with a neat centre parting, and he was wearing a grey pinstripe suit and a shirt in ivory silk. He walked in slowly, allowing everyone to get a good look.

They had just started on the goose when the bells of the nearby Sacré Coeur rang out. 'And to think,' said Perrette, 'what Mr Ruche was telling us about before dinner happened nearly four hundred years before the birth the bells are celebrating.' They began to discuss the session: the laws and methods at one's disposal to solve a problem. All of them were thinking about Grosrouvre and how he had become rich. He had admitted in his letter that he hadn't always followed the law. Perhaps he had been involved in trafficking something: jewels? gold? animals?

'But he said he didn't have blood on his hands', said Perrette.

'Yes', agreed Mr Ruche, 'I think he was simply letting me know that he didn't use *every* method at his disposal, unlike the gang who were after him. It's pretty clear that they would stop at nothing.

'The symbols of Communism are the hammer and sickle. Religious symbols include the cross and the crescent. Kings rule with the orb and the sceptre. What about the Greeks?'

'The ruler and the compass', everyone shouted together. They raised their glasses of champagne.

Sidney was woken by the toast, and Albert suggested they offer him a little champagne. 'For God's sake, no!' Mr Ruche grabbed the bottle, drew himself up in his chair and declaimed: ' "The bird from

India, which we call the parrot and which, it is said, can speak the language of man, cannot be silenced when it has drunk wine!" Aristotle, from *Stories of the Animals.*'

So Sidney didn't get his champagne, but Perrette gave him a plate of honey treats. 'So, the Greeks never did solve their three problems, then,' she said to Mr Ruche, 'a thousand years after they had first asked them, they still hadn't solved them!'

'You tell him, Mum!' cried Lea. 'They couldn't solve them because of the ruler and the compass.'

'I think you're being a little hasty', said Mr Ruche. 'OK, the Greeks didn't manage to solve their problems then, but the story isn't finished yet. Other mathematicians came after them. How do you know that one of them didn't manage to solve one of the problems with a ruler and a compass? Or even all three? Well?'

'So what are we supposed to do?' asked Lea. 'Do we break the rules like Hippias and Archytas and use methods that are "illegal"? You said the reason the Greeks wouldn't accept mechanical curves was because they introduced the concept of movement into maths. Well, aren't we in a similar position with Grosrouvre? I mean we haven't moved from here.'

Mr Ruche smiled.

'Sorry, Mr Ruche, I wasn't having a go at you,' said Lea, 'but it's true, isn't it? I'm just wondering if we can ever solve the mystery of Grosrouvre's death without going to Brazil.'

'There's another mystery which we haven't even mentioned', said Jonathan. 'Did Grosrouvre really prove Fermat's Last Theorem and Goldbach's conjecture?'

CHAPTER 12

The Great Glass Elevator

Mathematics in the Arab world; poetry and algebra & Omar Khayyām

Mr Ruche had never suffered from insomnia, but that night he woke up in the early hours. It was Grosrouvre's letters – he couldn't stop thinking about them. He was certain that, if he read between the lines, he would see what Grosrouvre was trying to tell him.

In the first letter Grosrouvre had listed a number of mathematicians. He claimed to have picked them at random, but had he? Or had he picked them deliberately, for some reason he intended his friend to discover? The only way Mr Ruche could find out was to study the mathematicians Grosrouvre had mentioned and try to figure out from their lives and their work something that would help him make sense of what had happened in Manaus.

Something Grosrouvre had written was bothering him. Mr Ruche sat up in bed, turned on the light and took out the neatly folded letters from the drawer of his bedside table. He immediately found the sentence he had been thinking about: 'If I remember, when we were young, when I kept secrets from you, you always managed to work them out.'

'That's simply not true,' thought Mr Ruche, 'Grosrouvre and I had many secrets from each other. What he's really saying is, "I am hiding something in this letter. Try to work out what it is".'

Mr Ruche's eyes lit up suddenly as he realized that Grosrouvre was not trying to hide something from him, but from others who might try to find out about his proofs.

'Just like Grosrouvre,' he thought, 'leaving it to other people to work things out!' As he folded the letter he noticed a phrase: 'I think I've said enough.' Mr Ruche smiled, happy to find himself caught up in the sort of game he and Grosrouvre had often played.

Now, from beyond the grave, Grosrouvre was sending him two messages:

'I have to hide certain things from you.'

'I have told you enough for you to work them out.'

Just like the texts written by the Pythagoreans, there was a hidden meaning reserved for the initiated, thought Mr Ruche. 'If I'm right, everything we need to answer the questions is in these letters. All we have to do is carefully follow the clues he gives, study the mathematicians he mentions.'

He glanced at the clock. It was 3.30 a.m. Mr Ruche shivered, though it wasn't cold. He put the letters back and turned off the light, but he couldn't get back to sleep. The first mathematicians Grosrouvre had mentioned in his letters were Persians: Omar al-Khayyām and al-Tūsī...

The next morning, Albert dropped Mr Ruche off on the Pont Sully, near the entrance to the Institut du Monde Arab – the Centre for Arab Studies – known by everyone as the IMA. He hardly recognized the area. Forty years ago there was a wine market here surrounded by a warren of little buildings and gardens with narrow paved streets lined with old trees. From across the street he was puzzled to notice that the surrounding buildings cast a strange reflection on the upper windows of the IMA. Looking carefully, he noticed that it was not a reflection, but silhouettes of buildings which had been printed into the glass. Mr Ruche rather admired the architect who preferred the reality of photography to the virtual world of reflections.

Since the library did not open until midday, Mr Ruche had time to look around. The building was woven out of metal and glass, but it seemed as though the Institute was built out of light itself. Sunlight flooded into it from all sides, the rays falling in a myriad of angles. The centrepiece was a huge atrium, with four glass lifts rising and falling in a silent dance. The only sound was a soft *ding* as they stopped at each floor.

At noon, Mr Ruche wheeled himself into one of the lifts. At the other side of the atrium, another lift had moved off at the same time. Looking across, it seemed like an air bubble in a column of water, carrying its passengers upward. It was magical. Mr Ruche was already planning to make one for himself – a lift for a new millennium.

The library extended over three floors. There were no stairs, just a simple spiral ramp alongside which were shelves crammed with books – Mr Ruche had never seen spiral bookshelves before.

He pushed his wheelchair onto the downward slope. Omar al-Khayyām's poetry was in aisle 8. He took several books and made his way to the huge, high-ceilinged reading room. Mr Ruche hung his coat on the back of his wheelchair and sat at one of the metal desks.

Unlike the National Library, the books here were directly accessible to the readers. You could simply browse the shelves and take down what you wanted. For the books higher up, which Mr Ruche could not reach, he asked a pretty brunette sitting near him to pass them to him, which she kindly did.

Omar al-Khayyām was not only a mathematician, but a poet. The first book that Mr Ruche opened was the famous *Rubāiyāt,* a poem written in quatrains. A note on the text explained that lines one, two and four were linked by rhyme, while line three was free.

> Ah, fill the Cup: – what boots it to repeat
> How Time is slipping underneath our Feet:
> Unborn To-morrow and dead Yesterday,
> Why fret about them if To-day be sweet!
>
> Another said – 'Why, ne'er a peevish Boy,
> Would break the Bowl from which he drank in Joy;
> Shall He that made the vessel in pure Love
> And Fancy, in an after Rage destroy?

The poem was full of a gentle insolence, a sweet provocation.

> Lo, some we loved, the loveliest and best
> That Time and Fate of all their Vintage prest,
> Have drunk their Cup a Round or two before,
> And one by one crept silently to Rest.

Mr Ruche put the book down on the metal desk, feeling suddenly sad. How long would it be before he crept silently to rest? But he shouldn't think about that now, he had a job to do.

Another quatrain, further down, reminded him of his true purpose here. It seemed almost as though it had been written by Grosrouvre to his friend:

> Then to the rolling Heav'n itself I cried,
> Asking, 'What Lamp had Destiny to guide
> Her little Children stumbling in the Dark?'
> And – 'A blind Understanding!' Heav'n replied.

If Grosrouvre had solved the equations, as he said he had, then he had overcome blind understanding and found the light. That must have made up for all the years of stumbling in the dark.

Mr Ruche closed the *Rubāiyāt* and opened a biography of Omar al-Khayyām. He was deeply engrossed in his reading when he heard a whirring sound. He looked around, but couldn't see anything. The sound continued and his eye was caught by the great glass façade. What he saw astonished him. A metal shutter, like the iris of an eye, surrounded each window and they were all closing slowly. It lasted no more than a minute, then the noise stopped – the shutters were almost closed.

The young woman who had helped Mr Ruche could not help but laugh at his astonishment. 'There are 27,000 of them exactly,' she said, explaining that there were 240 panes in the façade, each with more than a hundred shutters. A photoelectric cell linked to a computer regulated the light that flooded into the building. When the sun was too bright, the shutters closed, like a squinting eye. She was studying architecture and had come to see how the building worked. The young woman pointed out that each pane represented the classic elements of Arabic geometry. She showed the delighted Mr Ruche that the figures moved round in rotation, the architect having cleverly combined squares, circles, octagons and stars.

She went back to her reading, leaving Mr Ruche to his thoughts. He almost felt like a student again, meeting a stranger in a library. He opened the *Rubāiyāt* again and his eyes fell at random on a verse:

> The Worldly Hope men set their Hearts upon
> Turns Ashes – or it prospers; and anon,
> Like Snow upon the Desert's dusty Face
> Lighting a little Hour or two – is gone.

'I feel as though I've known Khayyām all my life,' thought Mr Ruche, 'even though he died long ago. I know how quickly knowledge fades – what men set their hearts upon. Though recently I've learned so much, thanks to Grosrouvre. I've also learned how much I don't know and what a pleasure it is to be ignorant while you can still learn.'

As he left the IMA, one phrase from the *Rubāiyāt* revolved inside his head: 'The Worldly Hope men set their Hearts upon/Turns Ashes – or it prospers'. What would become of the investigation Grosrouvre's letter had begun – would it prosper?

Back at the house, Mr Ruche had a short nap and woke up full of energy. He had the whole afternoon to himself: the twins were off skiing, Max was out somewhere, probably at the flea market, and Perrette was at the bookshop. He threw a rug around his shoulders and wheeled himself out of his room and across the courtyard. It was bitterly cold – it felt as though it might snow.

He pushed open the door to the Rainforest Library. It was dark and warm inside. He turned on a few of the spotlights, took the rug from around his shoulders, went to his desk, opened his notebook and began re-reading his notes. Then he wheeled himself across the library to Section 2, Mathematics in the Arab World.

He scanned the shelves: *The Book of Knowledge Concerning Plane Figures and Spheres* by three brothers called Banū Mūsā. The *Book of the Ingenious Workings and Mysterious Nature of Geometric Shapes* by Abū Nasr al-Fārabī. Three books by al-Karajī: the *Al-Badi*, the *Al-Fakhri* and *The Whole Science of Arithmetic*. There was the *Treatise on Shadows* by al-Bīrūnī, the *Illuminated Book of Arithmetic* by al-Samawa'l, *The Key to Arithmetic* by al-Kāshī...

Mr Ruche spotted the books by al-Khayyām and took them down, together with some books by al-Tūsī. There was something strange in these, in that al-Tūsī's first name was sometimes written as 'Sharaf al-Dīn' and sometimes as 'Nasīr al-Dīn'. He must have made a mistake in cataloguing. He checked his list of Arab mathematicians, only to find that there were two al-Tūsīs. Both were Persian: Sharaf was born in the twelfth century and died early in the thirteenth, and Nasīr was born in the thirteenth century. Which one did Grosrouvre mean when he mentioned al-Tūsī in his letter? Having two of them was not going to make his work any easier. Then he spotted a book by Thābit ibn Qurra called *A Short Work on Amicable Numbers*, which he quickly pulled down. He took out Grosrouvre's index card from the back and read his friend's spidery handwriting:

> The oldest copy of Euclid's *Elements* dates from the ninth century. Thābit ibn Qurra made a new copy some decades later. While Euclid ignored the 'friendly numbers' so dear to Pythagoreans, Thābit ibn Qurra devised a way of discovering pairs of amicable numbers and wrote what was to be the great theorem on the subject. The Greeks were aware of only a single pair of amicable numbers...

'220 and 284, I know, I know', Mr Ruche muttered to himself.

> ...Arab mathematicians would discover others: al-Fārisī discovered the pair 17,296 and 18,416, known as *Fermat's pair* because Pierre Fermat rediscovered them centuries later. Al-Yazdi discovered the pair 9,363,584 and 9,437,056, known as *Descartes's pair* because René Descartes rediscovered them centuries later.

Seeing Grosrouvre's handwriting upset him more than he expected. When had he written this card? Years ago, probably. Mr Ruche tried to imagine Grosrouvre in his house in Manaus – a young man still, leaning over his desk, working...He stopped, saddened. He couldn't imagine Grosrouvre in his house. He didn't know whether it was in the jungle or a leafy suburb. Was there a view of the great river Amazon itself, thundering past his windows? Mr Ruche could barely imagine what it would be like to live so near the equator. He had never liked hot weather and he hated it when it was humid. Remembering the chilly air of the courtyard made him feel better.

He remembered that Grosrouvre had mentioned Fermat in his note and mentioned the amicable numbers with which he had ended his last letter. 'What about us – are we "friends"? What is the sum of those things that define you or me? I think perhaps the time is coming when we will find out.'

'For you, old friend, the time has come and gone', thought Mr Ruche. 'But what about me?'

The door opened and Perrette came in. 'It's nice and warm in here!' She held back a smile as she placed an envelope on Mr Ruche's desk. 'The twins have sent us a letter.' She hesitated; she could feel that he was miles away. 'I'd better get back to the shop – I've got customers. I even had two people in this morning who I haven't seen for years, and they both bought lots of books to give as presents.'

'So you've had a lot of gift wrapping to do?' Mr Ruche asked, a note of fear in his voice. He hated gift wrapping.

'I love wrapping presents', Perrette said. 'When I was a kid, I used to do it all the time. I used to wrap anything I could find: matchboxes, a pair of shoes, string beans...I even used to make tiny presents with a single sugar lump inside using my mother's tweezers... Anyway, it's nice to see people giving each other books again.' She closed the door behind her.

Mr Ruche wheeled himself to his desk and tore open the envelope. He took out two photographs. In the first, Jon-and-Lea were standing at the top of a ski run, looking immaculate. On the back was the word 'before'. In the second, they were in a heap on the ground covered in snow, their skis and sticks everywhere. On the back, it said 'after'. Mr Ruche burst out laughing.

'Before and after...' He wheeled himself back to the bookcase to look for books by Omar al-Khayyām. Mr Ruche wondered if they had snow in Samarkand, where Omar al-Khayyām had completed his seminal work in algebra. He took down two, *Algebra* and *The Division of the Circle*, and went back to his desk. He slipped out the index cards and read the first:

> Al-Khayyām was the first to advance the idea of the polynomial. Early algebra consisted entirely of the study of equations. Al-Khayyām extended this to cover the study of polynomials. He performed the standard operations on them – addition, subtraction, multiplication and, particularly, division (he applied the Euclidean division of numbers to polynomials) and the cube root of polynomials.

Mr Ruche jotted an equation on his notepad to try to clarify the idea in his own mind. He muttered aloud as he scribbled: '$ax^2 + bx + c = 0$ is a second-degree equation. I can calculate its roots. But if I simply write $ax^2 + bx + c$, it's not an equation but a polynomial – a second-degree polynomial – and since there are three terms, it's a trinomial: a second-degree trinomial!' Suddenly, some of the mystery of the maths of his childhood seemed clearer. 'In that case, $ax + b$ is a first-degree binomial. I suppose a monomial would only have one term.'

He went back to the last page of al-Khayyām's *Algebra*, which ended with the words 'Finished at noon, on the first day of the week, the twenty-third day of Rabia in the year 600.'

Al-Khayyām established a complete classification of 1st, 2nd and 3rd degree equations. While al-Khwārizmī had principally dealt with 2nd degree equations, al-Khayyām dealt mainly with 3rd degree equations, of which he identified 25 separate types, based on the number of terms they contained. He resolved them using geometric processes.

N.b. Following the work of al-Khujandi, al-Khayyām confirmed that the equation $x^3 + y^3 = z^3$ (in today's notation) had no solution in whole numbers. He was not far from proposing Fermat's Last Theorem – in the 12th century!

Grosrouvre had mentioned al-Khwārizmī in a number of his index cards, so Mr Ruche thought he should investigate. He spent a couple more hours in the Rainforest Library researching al-Khwārizmī's work.

By the time he left, it was snowing heavily. Under his blanket, safe from the elements, was a book by al-Khwārizmī. When he got inside, he read the first sentences just before he fell asleep:

> Scientists of times past and of civilizations forgotten have always written their thoughts in books. They did this so that they might leave their wisdom to those who came after them. In this way, the search for truth goes on, for the secrets of science they uncovered and the light they brought to the dark corners of human knowledge would not be in vain. A man may discover something new and bequeath it to those who come after him. Another might discover something unknown to the ancients and so shed light on a new path, making future work easier for another. Still another may find errors in a book and try to rectify them without damning the author or taking glory for his corrections.

'A man may discover something new and bequeath it to those who come after him.' Grosrouvre should have read this, thought Mr Ruche, it might have cured him of his mania for secrecy.

Mr Ruche had to admit that, for the first time, he missed the twins while they were away. When they arrived back that evening, loaded with bags and buzzing with energy and enthusiasm, Mr Ruche felt a sudden warmth. He watched them as they crossed the living room and was surprised to see that Jon-and-Lea were limping – a result of the catastrophe immortalized in the 'before and after' photos they had sent. Their tanned healthy faces and the white skin where their goggles had blocked the sun made them look like climbers back from Everest, but Sidney recognized them immediately and squawked a greeting.

Jon-and-Lea clambered up the stairs to their attic rooms, where they tended to their war wounds. Lea rubbed foul-smelling ointment into Jonathan's bruised ankle, and he massaged her swollen knee. Then they settled down in their separate rooms, each with one foot raised on a cushion.

CHAPTER 13

The Story of Ø

The power of zero; Brahmagupta and al-Khwārizmī & a tea ceremony

The pale light of early January flooded through the windows of the screening room. Seated in the middle of the room, Mr Ruche began.

'A man is walking down the street looking for directions. He passes a stranger and asks him, "How do I get to X Street?" and the stranger replies, "If you don't know, you shouldn't go there."'

Everyone laughed.

'Well', he went on, 'algebra is exactly the opposite – if you don't know, you should go there.'

As he spoke these words, a black curtain was drawn across the windows, and Max appeared holding a lighter. He leaned over and lit a number of candles, each in a clay pot on a bed of sand. The sand was a precaution against fire, but it was also a bit of the desert brought to the Rainforest Library. In a corner, a small teapot was being heated on a brazier and beside it, on a copper tray that glowed in the candlelight, were glasses.

The room filled with the smell of incense, and there was a low sound of a lute playing. Jonathan closed his eyes and let himself drift. He felt like Lawrence of Arabia, riding a camel across the desert dunes, and felt calm. Suddenly, the lute faded and drums began to beat. Jonathan jumped up with a start. The drums seemed so close. Somewhere in the studio, someone was playing a tom-tom.

He opened his eyes and looked around, but he could not work out where the drumming was coming from. With a final dizzying patter the drumming stopped. The overture had finished, it was time for the main event. In his wheelchair, Mr Ruche nodded his thanks to the invisible musician and looked around admiringly at Max's decorations.

Jon-and-Lea sat up, eager to start the first session of the New Year.

'It all began on a day in AD 733, when a caravan which had travelled all the way from India came to the gates of *Madinat al-Salaam*, the City of Peace – what we call Baghdad – bearing gifts for Caliph al-Mansūr.

'Baghdad had been built in barely three years and, like Alexandria, was flanked by water on either side – the Tigris and Euphrates rivers. It was a cosmopolitan city, but while Alexandria had been constructed on a rectangular grid, Baghdad was circular. It was called the Round City. The city wall was a perfect geometric circle. At the centre of the circle stood a mosque and the Caliph's palace, and from here four great roads fanned out to the gates which were the only entrances to the city.

'A direct descendant of the prophet Muhammad, the Caliph was the Commander of the Faithful, giving him power over Muslims the world over. And by the end of the eighth century, there were many Muslims in the world. Islam had spread with astonishing rapidity. The Muslim Empire was the true successor to Alexandria and Rome. It stretched from the Pyrenees to the banks of the Indus and included the Iberian peninsula, the Maghreb, Libya, Egypt, Arabia, Syria, Turkey, Iraq, Iran, the Caucasus and the Punjab.

'To unify the peoples they recently converted to Islam, a common language was needed. That language was Arabic. From a language spoken by few, it grew and adapted to become capable of expressing ideas which the desert people who originally spoke it could not have imagined. It is a language structured to deal with abstract concepts. You might say it was made for algebra. The development of Arabic was made possible by books.

'In the district of Alexandria-Karkh in Baghdad was the biggest book market in the world. Manuscripts written on papyrus or parchment were brought from everywhere: Byzantium, Alexandria, Pergamum and Syracuse, from Antioch and Jerusalem. Here they were sold to buyers prepared to pay vast sums for them.

'The library of Hārūn al-Rashīd was a true successor to the Great Library of Alexandria, but it was his son, al-Ma'mūn, who founded the greatest institute of the Muslim Empire: the House of Wisdom – *Bayt al-Hikma*. The books that had flooded into Alexandria were, for the most part, written in Greek, whereas not one of the books in the Library of Baghdad in the ninth century was written in Arabic. Each one had to be translated. It was a massive undertaking!

'The team of translators who worked for the library of Baghdad was its greatest asset. There were hundreds of them from all over the world working on manuscripts in Greek, Sanskrit, Latin, Hebrew, Aramaic and Coptic.

'In the huge writing rooms, an army of scribes worked non-stop, and manuscripts, translated into Arabic, began to appear on the shelves of the library in the House of Wisdom. In no time at all, the Arab world assimilated a wealth of knowledge into its culture and traditions. For seven centuries, it was in the Arab world that the sciences prospered as this vast font of knowledge spread across the Muslim Empire.

'As knowledge spread, private libraries sprang up. The most prestigious, owned by the mathematician al-Kindī, was much fought over after his death. In the end, it was given to the Banū-Mūsā brothers, Muhammad, Ahmed and al-Hasan. The brothers were an institution in their own right: they had their own translators, whom they sent abroad at great expense to bring back rare manuscripts.'

'Mr Ruche?' Jonathan asked innocently, 'doesn't that remind you of someone?'

'Of course it does – Grosrouvre. But in his case, the books came to him. The library of Baghdad was subsidized by the Caliphs, who were passionate about the arts and sciences. They began a hunt for manuscripts on a scale unheard of since Ptolemy a thousand years before. After al-Mansūr, who had accepted the gift from the Indian travellers, came Hārūn al-Rashīd. His son, al-Ma'mūn, was an extraordinary man – a rationalist and a disciple of Aristotle who was the heart and soul of the House of Wisdom.

'When his troops defeated the Byzantine army, he proposed to exchange the Byzantine prisoners of war for manuscripts. A thousand Christian solders were freed in return for a dozen rare manuscripts – the pride of the Byzantine collection.

'But let's go back. I said that it all began one day in the year 733, when a caravan which had travelled all the way from India came to the gate of Baghdad bearing gifts for Caliph al-Mansūr. They were met by the Caliph in the great reception hall of the Palace. Al-Mansūr wore the coat of the Prophet and a pair of magnificent red boots, and carried a cane, a sabre and a seal: the symbols of his office.

'Among the sumptuous gifts they had brought was the *Brāhmasphutasiddhāntha* – a treatise on astronomy written a century before by Brahmagupta. It was immediately translated into Arabic and became famous as the *Zīj al Sind-hind*. It revolutionized science in the Arab world through ten little symbols that every one of you knows well. The ten numbers we use to calculate: 1, 2, 3,... up to 9, not forgetting the last – "zero".

'The man who brought the book knew the symbols and had used them for years to calculate. He had changed them so often on the journey to the Round City that everyone in the caravan knew them by heart. At night, around the campfire, one of them would begin and the others would start to sing out the numbers in the dark.'

In the darkness of the library, Sidney's voice rang out: '*Eka, dva, tri, catur, panca, sat, sapta, asta, nava.*'

'What about zero?' asked Lea.

'*Sūnya*', said Mr Ruche, who had kept the honour of introducing 'zero' for himself.

There was a long drum-roll.

'*Sūnya* means "empty" in Sanskrit. Zero was represented by a little circle. "Empty" translated into Arabic becomes *sifr*, which in Latin becomes *zephirum*, which in turn becomes *zephiro* in Italian, and from there it's a short step to *zero*.'

Mr Ruche stopped, suddenly remembering the paper Grosrouvre had published on the number zero nearly fifty years before. The essay that had given them their nickname 'Being and Nothingness'.

'These ten numbers were the most important elements of a system which allowed numbers to be written and calculations to be done. It is called the "positional system of decimal numbering". Undoubtedly it was one of the most important discoveries in human history.' Mr Ruche paused for a moment. 'Why "positional"? Well, if you lot aren't going to ask me, I'll have to tell you. Are you all asleep?'

'No!' said Lea, 'I was just so wrapped up in what you were saying.'

The twins really did seem to be interested. Mr Ruche had noticed that numbers fascinated most people – sometimes to the point of mania. In his many years in the bookshop, Mr Ruche had met many numerologists, and now he avoided them like the plague. In the short time he had been studying it, Mr Ruche had become as passionate in his love of arithmetic as he was in his loathing of numerology. The wonder of numbers was in the numbers themselves. There was no need to look for some quasi-magical significance in numbers, he thought. Their magic could be found everywhere: the most recent example he had discovered was the existence of 'twin' primes. Two prime numbers are 'twin' if they could not be closer to each other, i.e. if the difference between them is 2. 17 and 19 are twin primes, as are 1,000,000,000,061 and 1,000,000,000,063. No one knows if there are an infinite number of twin primes but they are extremely rare.

The embers of the brazier sparked and glowed. Mr Ruche dragged himself from his reverie and answered the question he had asked some minutes before: 'Almost all cultures have developed a system for writing numbers. Some are efficient; others, like Roman numerals, cumbersome. In most of them, the value of a number is the same regardless of where it appears. For example, X in Roman numerals equals ten, regardless of where it appears, so XXX equals thirty: ten plus ten plus ten.

'Positional number systems work differently: the value of the number depends on its position. "1" represents one, ten or one hundred, depending on whether it appears last, second from last or third from last. The 1 in 1000 is worth more than the three 9's in 999.

'This Indian number system did something remarkable: with just a handful of symbols – exactly the same number as the fingers on both hands – they managed to represent all the numbers in the world. Indian mathematics was far in advance of that of other civilizations. If ever an invention had universal implications, this is it: nowadays, everyone in the world uses this number system.'

A voice came from the darkness: 'My friend, as an Arab I think you are robbing our numbers from us.' Everyone was speechless. It was Mr Habibi from the corner shop, who came out of the shadows. It was he who had played the lute and the drums. 'These numbers were invented by Arabs, and the zero too!'

'I'm sorry Habibi, that's what I thought too, but we were wrong – Indians invented the numbers we use nowadays. You can't rewrite history.'

'Then you tell me why everyone calls them "Arabic numerals".'

'When the numbers came to Baghdad', explained Mr Ruche, 'the Arabs called them "Indian shapes". A mathematician, one of the members of the House of Wisdom, wrote a thesis to explain the system and how it worked. It was through him that the Arabs discovered these numbers. Centuries later, the manuscript was translated into Latin and it was this translation which introduced the system to France, Italy and Germany, and from there, throughout the world. Since Christians learned of the numbers from the Arabs, they called them *Arabic numerals* and decided that the number zero was an Arabic invention. That is why everyone now calls them "Arabic numerals" rather than "Indian numerals".'

'This thing you tell me makes me sad, Mr Ruche', said Mr Habibi. 'It is like you are telling me that couscous is invented by the Irish.'

Max, who hadn't quite followed the exchange, noticed that Mr Habibi was sad. He decided to cheer everyone up, and carried the tea tray into the middle of the room and asked Mr Habibi if he would pour. Habibi got up, went to the brazier and picked up the teapot. He took a glass and held it close to the ground, lifted the teapot as high as he could and poured the boiling hot mint tea directly into the glass. As he poured, he moved the pot and the glass at dizzying speed. Not a single drop was spilled.

Everyone gathered in a circle around the tray. They drank the sweet mint tea and ate from the box of fresh dates that Mr Habibi had brought back from the village he came from in Algeria. When the last date was gone and the last drop of mint tea finished, Mr Ruche said quietly, 'There's no need to be sad, Habibi. The Arabs may not have invented numbers, but they did invent algebra.'

'Thales was the first Greek mathematician. The first Arab mathematician was a man named al-Khwārizmī.'

Mr Ruche stumbled over the name, so Mr Habibi kindly gave a demonstration of how it should be pronounced, and Mr Ruche tried again. This time, he gave the mathematician's full name: Abū

Ja'far Muhammad ibn Mūsā al-Khwārizmī. He got a round of applause. 'The name means – tell me if I'm right, Habibi – that he is the son of someone called Mūsā who comes from Khwārizm!'

Mr Habibi nodded. Mr Ruche handed him a book and asked him to read the title. Mr Habibi held it respectfully, almost fearfully, then carefully read the words across the front cover: *'Hisāb al-Jabr wa'l-Muqābala'*. He handed the book back to Mr Ruche, who read the opening lines: 'For the calculus of al-Jabr and al-Muqābala, I have composed a short work which succinctly captures the subtle glories of that science. In this, I owe much thanks to Ma'mūn, the Guardian of the Faithful. He is a prince who brings men together, helps and protects them and encourages them to make the darkness light, the complex, simple.'

Mr Ruche repeated the last sentence: ' "To make the darkness light, the complex, simple." It's not so much a system as a philosophy. This book is one of the most famous in the history of mathematics. Throughout these pages', he said, leafing carefully through the book, 'a new discipline, a completely new system was born: algebra. The word itself comes from the name *al-Jabr*.'

'*Al-Jabr* means a bone-setter', said Mr Habibi. 'Where I come from, if you have a *douar*, if you break something, you go to *al-Jabr* and he twist this way – Ow! – and that way, and set the bone back in place. Yes, *Jabr* is someone who fixes something that is broken.'

Mr Ruche jumped in, 'In the story of *Don Quixote* there is an "algebrist" – a bone-setter. I never understood why he was called an algebrist before. Cervantes must have learned the word from the Spanish Moors.'

'Wasn't there another word in the title, after *al-Jabr*?' asked Lea, careful not to try to pronounce it.

'Yes: *al-Muqābala* – this is when two things are put against each other', said Mr Habibi. 'How do you say this?'

'Confrontation?' asked Mr Ruche.

Lea laughed. '"A Thesis on the Calculus of Bone-setting and Confrontation." When I tell my maths teacher we've been learning to set bones, I'm sure he'll be impressed. If he argues, I'll send him to talk to you, Mr Habibi.'

'Let him come, let him come', said Mr Habibi.

'In algebra – a little like bone-setting – you move things from left to right, trying to find what is missing', said Mr Ruche. 'This is how al-Khwārizmī explains it: "That which I am searching for, I begin by naming. But since I do not know what it is; since it is this that I seek, I simply call it the *thing*." This is the unknown – what mathematicians might now call x or y. By naming the *thing*, as he calls it – though he does not have a value for it – he can work with it as he does with other numbers. His strategy is to try to calculate an unknown value by treating it as though it were known. He adds to it, subtracts from it, multiplies and divides, but in doing so he is trying to find out what it is. Discovering the unknown is the magic of algebra.

'You won't find equations you would recognize in this book. There are no plus or minus signs, and variables are not written as x. These symbols came much later. Al-Khwārizmī's equations are written out in words. There is something else that is strange: the Arabs did not recognize negative numbers. All the terms preceded by a minus in the equations disappear – they were called *naquis*, which means "amputated". Al-Khwārizmī deals only with positive numbers: whole numbers and fractions. Fractions are "broken numbers" – the word comes from the Latin translation of the Arab word *kasr*, which means "broken".'

'With all these broken numbers and amputated numbers', said Jonathan, 'it's hardly surprising they needed a bone-setter. Algebra is starting to sound like a casualty department.'

'You don't know how right you are', said Mr Ruche. 'There was another set of numbers he didn't acknowledge: irrational numbers, which he called *assam*. Do you know what that means? It means "deaf". Why? Because irrational numbers cannot be expressed using numbers.' Mr Ruche leafed through his notes, and read: 'Étienne Condillac, the French philosopher, describes it like this: "When we have no exact measure for a number, we say it is deaf, for it escapes us, like a dull sound that one barely hears."

'What about the word "root" ', asked Mr Ruche, 'where do you think that comes from?'

'From the root of a tree?' suggested Max.

'Exactly. What is the cube root of a?'

'A number which, cubed, equals a', said Max, pleased that Mr Ruche had asked him.

'Exactly. It is a number that has to be "extracted" because it is buried like the roots of a tree. When you've "extracted" it, you can "raise" it to the power of 3, or "cube" it. It's beautiful, isn't it?

'Equations outline not a specific problem, but a whole class of similar problems. For example, "something added to one number equals a second number", the problem being to find out what the "something" is if both numbers are known.'

'First-degree equation', said Jonathan.

'Al-Khwārizmī's speciality was second-degree equations', said Mr Ruche. 'He found six types of first- and second-degree equation:*

"squared numbers equal roots"	$ax^2 = bx$
"squared numbers equal numbers"	$ax^2 = b$
"roots equal numbers"	$ax = b$
"squared numbers plus roots equal numbers"	$ax^2 + bx = c$
"squared numbers plus numbers equal roots"	$ax^2 + b = cx$
"squared numbers equal roots plus numbers"	$ax^2 = bx + c$

Mr Ruche was not explaining from memory, but from the notes he had made from Grosrouvre's index cards. 'The word "equation" comes from the word "equal". Without the idea of equality, there would be no such thing as maths.'

'And no French Revolution!' said Lea.

'You believe that the French Revolution made people more equal?' asked Mr Ruche.

'We're still young, we're allowed to be idealistic', said Jonathan.

Mr Ruche smiled and nodded. 'An equation is like a pair of scales – it has two sides, and equality involves keeping them balanced.' He held out his hands, palms up, at the same height. 'If you put something on one side...' Max mimed putting something into Mr Ruche's right hand. His right hand fell, and his left hand rose; '...then the balance is broken.' Mr Ruche put his hands back as they were. 'If you subtract something from one...' Max pretended to remove something from Mr Ruche's left hand, which rose as his right hand fell; '...the balance is broken.

'I don't know if you remember – it was before you went on holiday – but Euclid talked about equality in many of his axioms.'

*The modern forms of the equations are given to the right of al-Khwārizmī's descriptions.

'If equals are added to equals, the wholes are equal', Lea sang out, imitating Sidney.

'If equals are subtracted from equals, the remainders are equal', said Jonathan, imitating Max.

'Well, an equation simply denotes equality between two expressions, at least one of which contains an unknown. Do you know, I got to be eighty-something before I finally understood that?' said Mr Ruche.

'Well, at least that means if we don't understand, we've got sixty-odd years to work it out', said Lea. 'And if we do, we're way ahead of you!'

'Equality is verified when equations are resolved', Mr Ruche announced.

'If they can be resolved', said Lea.

'And once the equation has been resolved and the unknown replaced with the value we have found, the equation becomes a statement of equality.'

'As long as we didn't get it wrong', said Lea, 'because if there's a mistake...'

'Then it won't be equal. In fact, that's exactly how we find out whether we've made a mistake', Mr Ruche jumped in.

'So, if I say "$2 + 2 = 4$" ', said Max, 'and "$2 + x = 4$" is an equation, am I ahead of you?'

'I'd say you're half a lifetime ahead of Mr Ruche!' said Lea.

Max smiled proudly. 'It's the other half that's going to be hard', he said quietly to himself.

Sidney flew down from his perch and landed on Max's left shoulder. Max let his shoulder drop until he looked like a hunchback, and said in a Quasimodo voice, 'The balance...the balance is broken!'

Mr Ruche turned out the lights in the screening room. The children were out in the courtyard helping Mr Habibi to carry his instruments. Mr Ruche took something from his pocket, something he seemed to have forgotten was there. He called to the children. Max didn't hear, and Jonathan was weighed down, but Lea ran back. He handed her an envelope. 'That's for you and your brothers,' he said.

'Thanks', said Lea, assuming it was a Christmas card. She could not have been more wrong.

Mr Ruche wheeled himself back to his bedroom and started his nightly ritual. He rolled the chair to the bed, putting down the side of the wheelchair, grabbed the armrest and hoisted himself from chair to bed. He took a deep breath and pulled his legs onto the bed. He took off his boots and let them fall to the floor with a dull thud.

As he lay back on his four-poster, Mr Ruche thought soberly that he had never found a bone-setter to put his dislocated body back together after his fall in the bookshop. Of course, he hadn't needed to be a *naquis*; nothing was amputated, he was like a fraction – broken. Just before he nodded off, he smiled as he remembered that the four-poster was invented in Baghdad too.

The following day, Lea did exactly what she had said she would, and told her maths teacher about bone-setting. The effect was startling, and an argument erupted in class C113. Two class swots accused her of bringing algebra into disrepute by comparing it to some barbarous medical practice. Lea was thrilled and listened gleefully as they dug themselves in deeper.

At lunchtime, she had arranged to meet her brothers in a café on the rue Lepic. Though he didn't say so, Max was proud to be invited out by his older brother and sister. Lea showed them the envelope that Mr Ruche had given her. Inside was a card with a riddle written on it:

> Perrette Liard said she had '2 + 1 children'. A pair of twins and one single child. The sum of their ages is 43 and the difference between them 5 years. How old are the Liard children?

Jonathan and Max looked at her, stunned, and burst out laughing. Max shook his head, took out a pen and a piece of paper and pushed it across the table. Lea took the pen.

'There are three Liard children', she said, 'but only two ages, OK? So it's an equation with two unknowns. Not bad. The first unknown is my age and Jon's, which are the same.'

'I'm two and a half minutes older', said Jonathan.

'Nit-picker', said Lea, contemptuously. 'Anyway, we'll call it x.'

'Exactly. The second unknown is Max's age, which we'll call y. The first thing we know is that the sum of the ages is 43. So?'

'So, $2x + y = 43$', said Max.

'OK, and the second thing we know is that the difference between the ages is 5, which means?'

'$x - y = 5$', said Jonathan confidently.

Lea wrote the equations out, one above the other:

$$2x + y = 43$$
$$x - y = 5$$

and said, 'Two equations, each with two unknowns, and now all I have to do is a bit of bone-setting and confrontation.' As she spoke, she scribbled notes on the paper:

$$x = y + 5$$
$$\text{Therefore } 2(y + 5) + y = 43$$
$$\text{If we multiply it out we find that } 2y + 10 + y = 43$$

'Now, if we take away 10 from each side we get $3y = 33$.'

'Which means Max is 11. Which he is!'

Max nodded, impressed.

'It also means', said Lea, 'that $x = 11 + 5$, meaning that Jonathan and I are 16!' She grabbed Jon's head and forced him to nod.

As they ate their lunch, Max looked a little worried. Eventually, he spoke up: 'There's something I don't understand, Lea: why did you write $x - y = 5$?'

'Because the difference between your age and mine is 5, of course.'

'Yes, but you're assuming that the twins are older than the other kid!'

'But they are!' said Lea.

'Maybe, but how do we know that? It's not part of the riddle. How do we know that the one kid isn't older than the twins?'

Lea looked at Jonathan. 'He's right, you know.' Lea ran her hand through her hair. 'Thanks for nothing, bro!'

Max laughed.

'What difference does it make?' said Jonathan.

'What difference does it make? I'll show you what difference!' Lea started her calculations again as her brothers watched. At last, she said, 'That would make Max seventeen and a half and you and me, brother, we'd be twelve and a half!'

'Yes, oh yes!' chortled Max.

Mr Ruche wasn't at the bookshop when they got back from school. Perrette said they would find him in Mr Habibi's shop. When they got there, Lea handed him the piece of paper triumphantly and told him how they had solved his equation. Then she told him about the other possible solution. He was surprised and a little embarrassed. A second solution hadn't even occurred to him.

That night, under the skylight of Jonathan's attic room, the twins pored over the map of South America, tracing the path of the river. At its source, among the peaks of the high Andes, the Amazon was less than 100 miles from the Pacific. Rather than run straight to the ocean, the river curved and meandered for 4000 miles to the Atlantic. At first, it fell quickly with spectacular waterfalls and cataracts. After that, it wound across the plains falling less than an inch for every mile.

'So we're definitely going?' Lea asked.

'Promise.'

The idea had come to them before Christmas. They would leave after their exams. Jonathan took a sheaf of brochures and guidebooks out of his briefcase. He looked again at the huge map of the Amazon basin and read out the names of towns along the river. Lea closed her eyes and listened. She could already imagine herself stretched on a comfortable hammock, on a barge sailing slowly down to Belem, while around her she could hear the humming of native songs.

Jonathan picked up one of the guidebooks and continued reading. As she listened, Lea felt fascinated and horrified by the history of the rainforest. There seemed to be no one left there to stand up to the big companies: the unions had been crushed and the native peoples enslaved, tortured and sometimes killed. Every day, thousands of acres of forest were burned to clear the land.

Jonathan read on: '"The Amazon is the garden of the world – though it is not the Garden of Eden. It is a marriage between heaven and hell. It is vast, so vast that it is difficult to imagine – it contains more than 15 per cent of all the world's vegetation. Trees tower over the undergrowth like natural skyscrapers, some of them more than 300 feet tall."'

'Here's something that will give you nightmares', said Jonathan. '"A single tree can contain up to 1500 species of insect!"' Lea shuddered. As her brother knew, she could hardly bear to be in a room with a spider. Lots of insect repellent would be needed, she decided. But she was willing to live dangerously.

CHAPTER 14

The Fifth Postulate

A gangster in Tokyo;
Menelaus and Nasīr al-Dīn al-Tūsī;
trigonometry explained & a fortunate barrow

'Books do not bring the dead to life', read Mr Ruche. 'They cannot make a madman sane, nor a foolish man wise. Books sharpen the mind; bring it to life, quench its thirst for knowledge. But for he who would know everything, it would be better that his family take him to a doctor, for this can come only from a troubled mind.

'Dumb when you wish it to be silent, eloquent when you wish it to speak, a book can teach you more in a month than the wisest men can in a lifetime and without a debt owed for knowledge gained. Books make a man free, deliver him from dealings with the odious, the stupid and those who cannot understand. A book is obedient, whether a man travel, or whether he be still. If you are disgraced, a book will serve you no less well. If others turn against you, a book will never turn its back on you. Sometimes a book may even be greater than he who wrote it.'

Dawn had just broken when Mr Ruche wheeled himself across the courtyard. It was snowing less heavily now, but it was bitterly cold. Though he had tried, he had not been able to sleep. He had so much still to learn, so many mathematicians still to study if he was to discover what Grosrouvre had been trying to tell him.

'If I'm right', he thought, 'there's a connection between the mathematicians he mentions and something in his own life. Something he knew I would be able to make sense of.'

Mr Ruche found his desk exactly as he had left it – in a mess. Sheaves of paper with calculations crossed out, a stained teacup, last year's newspapers and the 'before and after' photos of Jon-and-Lea.

The next mathematician on his list was al-Tūsī, but first he would have to find out which of the two Grosrouvre had meant: Sharaf or Nasīr. Mr Ruche wheeled himself along the shelves and took down *Equations*, a book of algebra by Sharaf al-Tūsī. Grosrouvre's index card began with the words:

Sharaf is the inheritor of the work of Omar al-Khayyām...

'That settles it', thought Mr Ruche, 'this must be the mathematician Grosrouvre intended.' None the less, Mr Ruche decided to have a look at the work of Nasīr al-Tūsī. He took down a book by Nasīr called *Collection of Arithmetic, Composed Using a Board and Dust** and another called *The Mysteries of the Secant Explained*.

According to Grosrouvre's index card, Nasīr al-Tūsī was the mathematician who had introduced trigonometry as a field of mathematics. Previously, it was used solely as a tool for charting the position of the stars and the movement of the planets. Through al-Tūsī, the geometry of circles and spheres became a separate discipline within mathematics.

Trigonometry is essential in calculating the movement of the stars and planets. It is not surprising, therefore, that Nasīr al-Dīn al-Tūsī's two important predecessors were both astronomers: Hipparchus and Ptolemy. There were two important mathematicians too: Theodosius and Menelaus.

A century after Euclid's plane geometry, Theodosius and, latterly, Menelaus, in *Sphaerica*, developed the geometry of the sphere. Menelaus discovered many of the properties of geometric figures on a sphere, in particular that the sum of the angles of spherical triangles is greater than 180°.

Greater than 180°? Mr Ruche was surprised. Surely the sum of the angles of a triangle – even on the surface of a sphere – was always *equal* to 180°. But according to Grosrouvre's card, that applied only to triangles in plane geometry. Mr Ruche had never really thought about the fact that there might be any other form of geometry.

'The sum of the angles of a triangle is equal to 180°', thought Mr

*Indian mathematicians and the Arabs who later followed them wrote directly on the ground, or on sand, or sometimes on boards covered with flour or sand which they carried with them. Numbers written in this way were often known as 'dust numbers'.

Ruche. 'This is something I've heard all my life. Something everyone accepts as an absolute truth – now it seems that it is only conditionally true.' Yes, it applied to all triangles, on condition that they are on a flat plane. The difference between flat and curved planes was explained by Menelaus: 'A triangle spread onto the skin of an orange would be "bigger" than the same triangle placed onto the leaf of an orange tree.'

This was the most important function of mathematics, he decided, to state precisely in which cases, under what conditions and subject to which hypotheses a statement is true. Grosrouvre's index card had reminded Mr Ruche how valuable mathematics could be as a reminder of the dangers of absolutism.

He picked up the index card on Nasīr al-Dīn al-Tūsī again. He had always hated trigonometry at school. All the talk of chords and arcs, sines and cosines had never quite made sense. It seemed to be a succession of words and formulas to be learned by heart to perform complicated calculations he didn't understand. He now was beginning to understand that it was really about the relationships between lines and circles. He made himself a sketch:

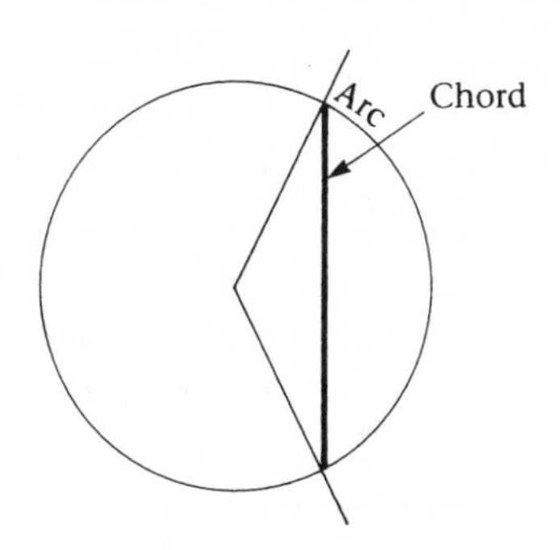

An *arc* and a *chord* are defined by each other. An *arc* is a segment of the circumference of a *circle*, and a *chord* is the line that connects the end points of the arc. It looked like an archer's bow. In a bow, the cord is held straight by the pressure of the wood, and the bow is curved by the constriction of the cord.

Trigonometry had first been used to establish astronomical tables which made it easier to track the movements of the planets. The first astronomical tables established the relationship between the length of a chord and the different values for the arc. Grosrouvre had added a short note:

> The table of chords is the first example of a *function* in the history of maths. It was at this time that the Greeks took to dividing the circle into 360 degrees.

Later, in India, mathematicians replaced the *chord table* with a *sine table,* which was easier to manipulate: the *sine* simply being half of the chord.* Mr Ruche made another sketch:

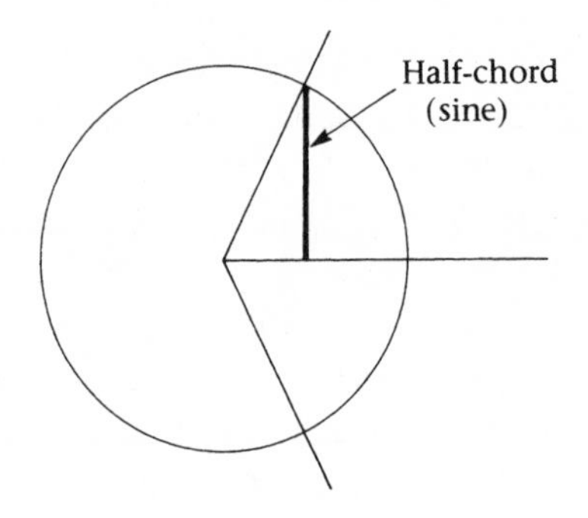

He went back to the index card:

> The precision of any astrological calculation depended on the accuracy of the sine tables, which related to one of the three great problems: trisecting an angle. It was al-Khwārizmī who first drew up a sine table. Shortly after, Habash al-Hāsib discovered the tangent. Al-Hāsib means 'the calculator', and the tangent is the best tool for calculating the height of an object.
>
> N.b. The Great Pyramid of Cheops could have been measured directly using a table of tangents – unfortunately, Thales did not have one.

Now Mr Ruche was back on familiar ground: sine, cosine and tangent. He picked up a pencil, took a compass and a ruler, and quickly sketched as much as he could remember from his schooldays:

*Trigonometric functions like sine were originally lengths, calculated for a circle with a radius of a certain value (120 units was often chosen, because 120 has many divisors, making calculation simpler). It was much later that mathematicians began to divide the sine length by the radius – the hypotenuse of the right-angled triangle – to give a pure number.

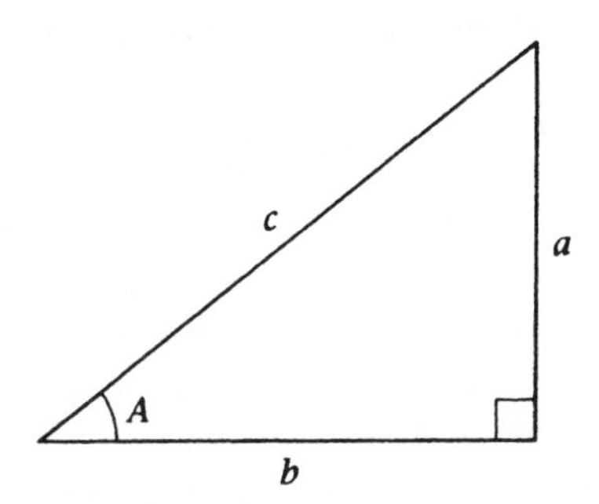

Sine: $\sin A = \frac{a}{c}$

Cosine: $\cos A = \frac{b}{c}$

Tangent: $\tan A = \frac{a}{b}$

Cotangent: $\cot A = \frac{b}{a}$

'To establish as complete a table as possible', Grosrouvre had added, 'Arab mathemeticians had to develop formulas' – one which Mr Ruche, like every other child, had learned at school:

$$\cos(A + B) = \cos A \times \cos B - \sin A \times \sin B$$
$$\sin(A + B) = \sin A \times \cos B + \sin B \times \cos A$$

So if we have two angles, A and B, and we know the sine and the cosine of those angles, we can calculate the sine and the cosine of the angle $(A + B)$ or of $(A - B)$. That was what the formulas Mr Ruche had learned were for: completing the exhaustive tables with only a small number of known values.

Mr Ruche closed the book, happy that he finally had some sense of what trigonometry was all about, but disappointed that he had found nothing to connect the work of Nasīr al-Tūsī with Omar al-Khayyām. Except perhaps that the first studied algebra and the second trigonometry, and only one discipline linked the two: geometry.

He took down *A Commentary on Difficulties with Certain Postulates in Euclid*, by al-Khayyām. All Euclid's postulates concerned geometry, he remembered. He could not find anything on geometry by Nasīr al-Tūsī. Could it be that the two had nothing in common in mathematical work? Mr Ruche found this hard to believe. If his theory was right, the book he was looking for must be in the Rainforest Library. He wheeled himself slowly along the bookshelves a second time. Then his eye fell on a book with a strange title: *A Short Work Quelling Doubts About Parallel Lines*, by Nasīr al-Tūsī – a book of geometry.

He brought the books back to the desk, opened al-Khayyām's and took out Grosrouvre's card:

> The work deals mainly with Euclid's fifth postulate, which defines parallel lines. It had bothered mathematicians since Euclid had first posed it in that it was phrased more like a theorem than a postulate. On the other hand, the fifth postulate was crucial to mathematics: without it, there was no proof of Pythagoras' theorem, no proof that the sum of the angles in a triangle equals 180°. Mathematicians had spent much time trying to prove the fifth postulate using deduction from axioms and other postulates. Al-Khayyām affirmed that two lines perpendicular to a third cannot converge or diverge on both sides, and this prompted him to suggest a new definition: that lines were parallel if they were each perpendicular to a common third line. This definition has the advantage that it can be tested, but having to test it means that it is no longer a primary property. It becomes impossible to say whether two lines are parallel without referring to a third. I don't like it.

Mr Ruche closed al-Khayyām's book and opened the one by Nasīr al-Tūsī and took out the card:

> Nasīr al-Tūsī also tried to prove the fifth postulate. He argued that al-Khayyām was wrong, but makes the same mistake in his own proof. Nasīr's depends on the fact that a perpendicular line and an oblique line must intersect at some point:

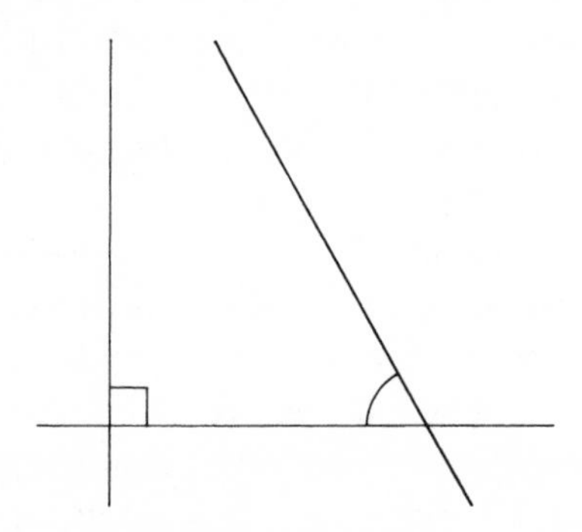

> Neither al-Khayyām or al-Tūsī, nor any other mathematician, has succeeded in proving the fifth postulate and the question remains open, like a thorn in the side of mathematics.

N.b. Nasīr al-Tūsī proposed a postulate which reads, 'If two straight lines on a plane diverge in one direction, they cannot converge in the other without intersecting'.

This was the link Mr Ruche had been looking for. Both al-Khayyām and al-Tūsī had tried to prove Euclid's fifth postulate; neither had succeeded. It didn't seem much. If he was to make progress, he would need to find out more about the lives of al-Khayyām and al-Tūsī. Maybe there was a stronger link there somewhere. He threw his rough sketches in the bin, put the teacup in his pocket so he could wash it under the tap in the courtyard, and took the photos of Jon-and-Lea and slipped them into an envelope. A piece of paper fell from the envelope as he did so and drifted across the library floor just out of reach. Mr Ruche took a long pair of pincers that he always kept under his wheelchair and picked it up. On it, Lea had written, 'Dear Mr Ruche, after quite a few falls, we have a postulate to submit to you: "For any given foot, there is only one possible ski and to that ski only one parallel ski." '

If Max had almost forgotten the men from whom he had rescued Sidney, they had certainly not forgotten him. Far away, one of them was about to find a clue to his identity.

The Tall Stocky Guy, as Max had called him, pushed past a group of schoolchildren on a bridge in the heart of Tokyo, walking quickly towards Shibuya Station to meet someone at the base of the Statue of the Dog. A client had told him the story of the statue. In the late 1920s a university professor would go to Shibuya Station every morning, followed by his dog, Hatchiko. A little while before his master was due home, the dog would come to the station and wait, and they would walk home together. Man and dog had followed the same routine for some years. One day, the professor did not come back. He had died in an accident earlier that day. But every evening Hatchiko still came to the station and waited for his master to return. Only when the last traveller disembarked would he turn back. For seven years, he came and waited. Then, in 1935 Hatchiko died and the citizens of Tokyo erected a statue to his loyalty and devotion. Arranging to meet someone at the Statue of the Dog meant you would wait as long as was necessary.

The man did not have to wait. His client was already there and their business was quickly concluded. It had been a good day. The Boss would be happy.

When night fell in Tokyo, the Tall Stocky Guy treated himself to a meal in a good restaurant. He ate sushi, washed down with some excellent sake, and then caught a taxi to Kabuki-cho, one of Tokyo's red-light districts. The taxi dropped him outside a karaoke bar. Inside, he was immediately enveloped in the warm sensual atmosphere of the bar. On a tiny stage, under a dim spotlight, a woman was singing.

The Tall Stocky Guy listened intently. Though he was a violent man, he was also sentimental and he liked love songs. When the singer finished, he applauded louder than anyone else.

The MC came over. 'You Français?' he asked. The man nodded. In fact he was Italian, but it was easier simply to agree. The MC handed him the microphone. 'Japanese like French song very much. You sing please?' The man shook his head. The MC turned to go, but as he did, he dropped the microphone. The man caught it before it hit the ground. The crowd laughed. This was an old trick – now that the man had the microphone in his hand he could hardly refuse to sing. The MC steered him towards the stage.

There was an expectant hush. The man sang '*Parlez-moi d'amour*' beautifully. Applause rang out as he finished and sat down. At the next table, two pretty young girls raised their glasses to him in salute. He raised his in return. They drank a toast to their good health and to love songs. One of them showed him a glossy magazine, and with a pronounced accent said, 'Paris, Paris!' She turned the pages and pointed to a photograph of the Louvre pyramid, under which was this:

高齢のフランス人学者は、建築家イェオ・ミン・
ペイの設計によるルーヴル美術館のガラス製ピラ
ミッドの高さを、古代ギリシアの数学者タレスの、
影を使う方式で測定する。

The man could not read a word of it, but something in the photo caught his eye. 'Good God!' he muttered to himself. In the middle of the photograph was a boy with a parrot on his shoulder. He stood up quickly, threw some money on the table, and, taking the magazine, ran out. It would still be mid-afternoon in Paris. There wasn't a moment to lose.

First thing the following morning he went into a nearby shopping arcade, found a shop with a photocopier and enlarged the photo. He asked if he could send a fax. The girl nodded and offered him a sheet of paper. He wrote quickly: 'Here's a photo of the kid. The parrot must still be in Paris. The ball's in your court – make sure you find them. Fast!' He sent the fax to his colleague in Paris. The Boss would be very pleased. The man adjusted his expensive suit and left the shop.

Back in Paris, Mr Ruche wheeled himself into the glass lift of the Centre for Arab Studies – the IMA. He had come back to find out more about Omar al-Khayyām's life. As soon as he reached the library, he pushed himself quickly up the spiral ramp into the main reading room. He chose some books from the shelves and settled himself at a desk. He worked quickly, hoping to get a lot done before he met Jon-and-Lea and Max in the café on the ninth floor at five o'clock. As he read, he took notes.

Omar al-Khayyām was born on 18 June 1048 in a small Persian village in the district of Khurasan. His father Ibrahim (the Arabic version of Abraham) made and sold tents, and when Omar became a poet he chose the name 'al-Khayyām', which means 'son of he who sells tents'. In a time when journeys took weeks or months and when many caravans of people travelled, it was a good business. Ibrahim sent his son to study at Nishapur. Omar quickly made friends, two in particular – Abdul Khassem and Hassan Sabbah. The three became inseparable. They studied together, talked together and, like students of every age, spent late nights at wild parties. It was after one such party that one of them proposed a pact: 'Let us swear our loyalty. We three are equal and alike and we should always be so. If one of us should rise to glory, he will help the others.' This they did.

The first to rise to glory was Abdul Khassem. Under the name Nizam u'l Mulk, he became Grand Vizier to Sultan Arp Arslan. Omar and Hassan Sabbah went to see him. Abdul had not forgotten their pact. He offered Omar a position of rank at court, but Omar declined. 'If you want to help me, give me the means to study for as long as I need', he said. Nizam gave him an allowance and had an observatory built for him in Ispahan.

Hassan accepted the post Nizam offered him. He was a cultured and intelligent man and soon found favour with the Sultan, but then he began to plot against Nizam, hoping to gain his position. As Grand Vizier, Nizam had many informants and much power. He learned of Hassan's plot and had him condemned to death. Omar interceded with the Sultan for his friend's life, and Hassan was exiled from the city instead. He travelled far, seeking a place where he might hide from Nizam's men, who had sworn they would have their revenge. Hassan had heard of a lost fortress south of the Caspian Sea, in the mountains of Elbruz. He decided to seek refuge there.

He left with a small group of companions. They made their way along treacherous paths on the edge of deep ravines. Then, high in the mountains, amid the ice and snow, Hassan saw the fortress of Alamut, perched like an eyrie on the summit. The fortress was surrounded by moats of freezing water – the only way in was across a drawbridge. The moment he saw it, Hassan decided that he would take it for himself, but since the fortress appeared impregnable, he could not do so by force.

Telling his companions to hide, he went forward and asked to be admitted to speak to the master of the fortress. His request was granted. Hassan spoke to the master: 'I have a bull's hide', he said, unfolding the hide as he did so, 'and I will give you 5000 pieces of gold if you will sell me as much land as can be measured by it.'

The master asked to see the gold, and Hassan showed it to him. He counted out the pieces: 5000. Convinced that he was dealing with a madman, the master accepted the bargain. 'Give me the gold and you may choose the land that pleases you immediately. Again, the drawbridge was lowered. Hassan went to a corner of the fortress and pointed at the ground. But, instead of placing the hide there, he cut the skin into narrow strips and tied them end to end to make a rope. Tying one end of the leather rope to a stake, Hassan walked around the fortress, trailing the rope behind him. It was just long enough to completely encircle the building. The fortress was his. The master had to leave, with his 5000 pieces of gold.

Once installed in the fortress, Hassan began to make strange changes.

Inside the sombre walls he created a paradise of beautiful gardens watered by clear streams, with flowers of all descriptions. He guarded this private paradise jealously. Only a few of his closest companions knew of its existence. It was a secret place, for which Hassan had a special plan.

From all across the Orient, Hassan chose two dozen young men famed for their energy and their skills as warriors. They were brought to Alamut where, for many months, they underwent rigorous training. On the day their training ended, Hassan had a great feast prepared in their honour. During the feast, Hassan drugged the soldiers with a secret herb, and they fell into a deep sleep. They were carried to the secret garden, where they woke the next day in paradise. The garden was full of beautiful girls who woke them with tender caresses.

They spent a whole day in this paradise, each moment filled with strange and wondrous pleasures; everything they had ever desired was here. When evening came, they feasted. Once again they were tricked into taking the mysterious herb. Hassan had the young men carried to their rooms.

When they woke in the morning, they were buzzing with excitement. They talked and talked about the paradise they had seen: how beautiful the women were, how gentle and loving. They talked about the breathtaking gardens, about the food and wine…it must all have been a dream.

Hassan assured them that it was not a dream, but a taste of paradise itself. They could return there, he said, but only if they died while carrying out his orders. They were to set out the next day.

Hassan had changed. He was now the Grand Master of a powerful religious sect called Ismailis. The Sultan and his Caliphs persecuted them for their beliefs. Hassan had declared war on the state and vowed to kill its leaders. His young warriors were his weapons. They were prepared to take any risk and were not afraid of death – which, they believed, was their passport to the paradise they had once tasted.

They were called *hashashins* – perhaps because of the herb which they took before each mission, hashish, or perhaps because they were mercenaries for Hassan. In time, the word *hashashin* became 'assassin'.

Mr Ruche's heart was pounding. He had set out to read the life story of a poet who liked wine and women, the 'Father of

Polynomials', a specialist in third-degree equations and a celebrated astronomer. Yet here he was, reading about a gang of assassins acting on the orders of a fanatical genius from a mighty fortress. Perhaps this was what Grosrouvre had been trying to tell him. He went back to his reading and note-taking.

> The *hashashins* did not fail Hassan. The Grand Vizier, Nizam, was found stabbed to death in his tent in the middle of the royal encampment. The *hashashin* sent by his childhood friend was executed immediately. As the executioner slit his throat, the man smiled.
>
> Hassan Sabbah died in his bed at Alamut many years later. He had not left the fortress since the day he had arrived. For a long time afterwards, people still spoke fearfully of the 'Old Man of the Mountain'.

Mr Ruche read for another hour, immersed in the story of the Ismailis. When he looked up at the clock, it was gone five. Mr Ruche wheeled himself quickly to the lift, which took him to the ninth floor. Lea, Jonathan and Max noticed immediately that he was excited. Mr Ruche ordered a mint tea and two small Lebanese pastries made with almonds and honey. The children had been expecting a maths lecture; instead, they were treated to a history of religion.

'Ismailism began around the seventh century, but was not always a murderous cult. After the death of Hassan, it became more pacifist. In Hebrew, Yishma-El means "God has heard". It was the name of Abraham's son by his slave Agar. "You will give birth to a son and you shall call him Ismail, said the Lord, for Yahweh has heard your distress." The doctrine of the Ismailis was to free the mind of obstacles and conditioning of any kind. They were scholars: the first philosophical and scientific encyclopedias were compiled by Ismailis, and the *Thousand and One Nights* are of Ismaili inspiration.'

'This is bad!' thought Lea. 'Mr Ruche has always been an atheist and now he's carrying on like a born-again Christian.'

Max, attentive as always, had lip-read everything.

As they drank their tea, Mr Ruche talked to them about what he had read that afternoon: about the three friends and Alamut. Jonathan interrupted: 'You started out saying you'd tell us about Omar al-Khayyām, but you keep going on about Hassan Sabbah.'

'Patience, Jonathan', said Mr Ruche. 'The only person Hassan admired was Omar al-Khayyām, his friend, the man who had saved his life. Moreover, Omar was a great scientist. On many occasions, Hassan asked Omar to come and live in Alamut. He refused, just as he had refused to work from within the Sultan's palace. He did, however, accept an offer to work on a new calendar. Omar al-Khayyām was one of the most reputed astronomers in the Arab world, thanks to his considerable talent, but also because of the fine observatory Nizam had built for him. For many years after, people would speak of "al-Khayyām's calendar". Omar was also an astrologer which is why we know exactly when he was born and when he died – something quite rare for the period. Once, he asked one of his disciples to ensure he was buried where the wind blew from the north and where the trees shed their blossoms twice a year. Many years later, when his student came back to Nishapur and discovered Omar had died, he asked where his mentor had been buried. He was taken to the grave. It was in a garden open to the north wind, by a small wall near which grew peach and pear trees and the grave was covered in a carpet of flowers from the trees' blossom.'

As the café closed, the twins went to spend the evening with friends and Max headed back to the rue Ravignan on foot. Mr Ruche wheeled himself out onto the ninth-floor balcony overlooking Notre Dame, and stayed there for a while thinking about al-Khayyām and how close he felt to him. 'Born on 18 June 1048', Mr Ruche remembered, 'died 4 December 1131.' Khayyām had died aged eighty-four, the same age as Grosrouvre.

Mr Ruche sat up in his wheelchair. 'And the same age as I am now!' he shouted into the cold Paris night. Mr Ruche was midway through his eighty-fourth year and at that moment he knew that nothing fatal would happen to him that year. He felt immortal, give or take a few years.

He went back into the reading room of the IMA. Albert would arrive to collect him soon, but he had a little time before the library closed. Time to read up on Nasīr al-Tūsī. He opened a biography and took notes as he read:

> Nasīr al-Dīn al-Tūsī was born in 1201 in a small village in north-west Iran called Tus, from which he took his name. His father was a celebrated

scientist, who, like Ibrahim, sent his son to Nishapur to the school where al-Khayyām had studied and where Nasīr read al-Khayyām's work. He was passionate about astronomy and dreamed of having an observatory at his disposal, like the one at Ispahan.

Two mathematicians: one, a lover of poetry, the other of religion. Nasīr al-Tūsī wrote *The Garden of True Faith*. This may explain why Nasīr al-Tūsī went to stay in Alamut, which was still run by the successors of Hassan Sabbah.

Mr Ruche could hardly believe his eyes. Nasīr al-Tūsī had stayed in Alamut. There was no doubt – he had to be the al-Tūsī Grosrouvre intended. 'Omar and Nasīr had both been in contact with the *hashashins*', thought Mr Ruche. 'That was what Grosrouvre was trying to tell me in the letter.' Thrilled with his discovery, Mr Ruche read on.

Nasīr al-Tūsī was delighted to discover the famous 'earthly paradise' at Alamut, but there was something that excited him more: the library that Hassan had created. Al-Tūsī spent most of his time there, and it was here that the Mongols came into his life. No one had been able to resist them. In less than fifty years, their troops had swept across Asia and Europe. After the death of Ghengis Khan in 1227, the Mongol Empire stretched from the Pacific coast of China to the Caspian Sea.

Lifting his head, Mr Ruche looked around him. The reading room was almost empty now.

The Mongol Empire succeeded those centred on Alexandria, Rome and Baghdad. By the year 1227, Peking, Moscow, Novgorod and Kiev had all fallen before the advancing Mongols and the army arrived at the gates of Vienna.

Khwārizm fell to the Mongols and the Aral Sea, Khurasan and Kurdistan, Iran and Iraq, Samarkand, Bukhara, Ispahan and Nishapur...In all this captured territory, only two places had held out against the Mongols: Baghdad, run by the Caliph, and Alamut, controlled by the *hashashins*.

Hūlāgū, one of the grandsons of the Great Khan, led his first military campaign against Alamut. The *hashashins* were hunted down and killed one by one. Only one thing remained: to take the fortress at the heart of the sect's power.

> One day in December 1256, Nasīr al-Tūsī heard cries. He rushed out of the library towards the battlements. Troops were advancing along the road. The men came on horseback, dragging behind them war machines that had toppled the walls of the mightiest cities in the world. Battle was about to begin.
>
> Alamut, the impregnable, was not taken; it surrendered. Al-Khayyām had lived through the founding of Alamut, Nasīr al-Tūsī through its destruction.
>
> The Grand Master who had succeeded Hassan Sabbah was beheaded, and the order given to raze the fortress. Not one stone was to be left upon another. When he came to the library, Hūlāgū stopped. He chose a learned man from among his company, pointed to a barrow nearby and said, 'I will give you one night to fill this barrow with books from the library. At dawn, everything that remains will be burned.' The man locked himself in the library and began to sift through the manuscripts. Why one book and not another? The hours passed.

As he read, Mr Ruche could feel his heart racing. He could feel the suffering of this man, forced to choose what would be saved. As a bookseller, he understood that in selecting some books he would condemn all the others to obscurity. Mr Ruche knew that for the rest of his life the man would have blamed himself for not being able to save them all. He read on:

> Outside in the snow, Nasīr al-Tūsī was watching. At dawn, he saw the man leave the library pushing the barrow, now full of manuscripts. One fell to the ground. Al-Tūsī moved to pick it up, but was stopped by a soldier. For seven days and seven nights the library burned. Hūlāgū spared the life of Nasīr al-Tūsī.

The Grand Master of the Alamut had not been able, as Grosrouvre had, to have the library transported to safety, Mr Ruche thought as he closed the book and prepared to leave.

Albert was waiting outside the IMA. They did not say much on the way back. Albert dropped him outside the bookshop, which was just closing. The lights were out and Perrette was pulling down the shutters. The moment she saw Mr Ruche, she knew that he needed to talk. They went back into the bookshop where she turned on a light and sat in her wicker chair. Mr Ruche told her everything he had learned that afternoon, and she listened in silence.

After a minute, she said, 'The library at Alamut was burned down, like Grosrouvre's house, and both al-Khayyām and al-Tūsī studied geometry and couldn't quite get their heads around…'

'The fifth postulate.'

'But apart from that, what else might the story tell us?'

Mr Ruche was silent.

'Let's start again from the beginning', said Perrette. 'Three friends meet while studying in Nishapur. The story tells us what happens to them as they get older. In your story, there are only two friends. But I don't know much about your life. Was there a time when there were three of you – you, Grosrouvre and someone else – all close friends? Maybe someone you haven't mentioned – that could be the link.'

Mr Ruche looked at her in surprise. 'Three of us?' He thought hard. 'No, I don't think so. Not at university. "Being and Nothingness", remember? And later, in the prison camp, there were a lot of men, but no one we were close to; there were still just two of us. We escaped together. Honestly, I can't think who the third person might be.'

'OK, well, we'll just have to think of something else.'

Mr Ruche was still deep in his past when Perrette suddenly surprised him by saying, 'What happened to the barrow?'

'The barrow? Oh, you mean the one the man used to move the manuscripts?'

Mr Ruche told her. 'After the fall of Alamut, Hūlāgū set his sights on Baghdad. He surrounded the city and laid siege. Resistance was useless. The Caliph sent emissaries to Hūlāgū. One of them was Nasīr al-Tūsī, who had come back to Baghdad after the Mongols had freed him.

'The Commander of the Faithful left the city to surrender to Hūlāgū, who told him to go back to the city with al-Tūsī and a handful of soldiers. Al-Tūsī reported the meeting between the Caliph and the Mongol Prince. Hūlāgū seized a golden plate and handed it to the Caliph. "Eat!" he ordered. "It is not food", protested the Caliph. "Why then have you kept it for yourself?" asked Hūlāgū. "Why have you not given it to your soldiers so they might better defend you?" Al-Tūsī reports that the Caliph was locked up with only his treasures for food. He died of hunger several days later.

'For the second time in his life, Nasīr al-Tūsī found himself in a city at the mercy of Hūlāgū. As at Alamut, there was a massacre and a hundred thousand people were slaughtered – a tenth of the population. For weeks, vast pyramids of skulls stood at every gate into the city marking the price to be paid for resisting the Khan.

'Hūlāgū asked Nasīr al-Tūsī to continue with his work. Nizam had built an observatory for al-Khayyām at Ispahan; a century later, in the town of Maragha, Hūlāgū built another, bigger still, for al-Tūsī.

'When he brought his possessions there, al-Tūsī had with him the barrow the learned man had used at Alamut. It was the thing he treasured most. Hūlāgū had given it to him.

'One by one, al-Tūsī catalogued the manuscripts that had been saved from Alamut in the great library of the observatory, which became one of the most important scientific institutions in the Middle East.

'The assassination of the Caliph resonated throughout the world. The capture of the city of the Commander of the Faithful was the end of the reign of the Caliphs, which had lasted five hundred years. And in Baghdad, Hūlāgū was succeeded by Tamerlain and the city was sacked once again. It was too much for the Round City, and for many centuries Baghdad was silent.'

CHAPTER 15

Secrets and Lies

Fibonacci's numbers; Tartaglia, Cardano and Ferrari; back to zero & the hunt for Max Liard

Mr Ruche would often take his books to Mr Habibi's shop in the afternoons. They would drink mint tea in the back room, and he would read while Mr Habibi worked on his accounts. Niccolò Tartaglia was the third mathematician on Grosrouvre's list, so Mr Ruche took two books from the Rainforest Library, *Quesiti e invenzioni diverse* and *General tratto,* and his notepad, and launched himself into the history of Italy in the sixteenth century.

> On the morning of 19 February 1512 the great church in the small Italian village of Brescia was crowded to capacity. These were not faithful coming to worship, but hundreds of women and children huddled together, terrified, trembling and hoping. Surely, they thought, they would be safe in God's house. Niccolò, his mother, his brother and his sister huddled near a pillar. The church was silent, all eyes fixed on the great door. Outside, the noise grew louder, closer. Inside, the silence was terrifying. The people held their breath.
>
> With a thundering roar, the door was broken down. A troop of soldiers piled through the yawning gap. Brandishing their swords, they urged their horses through the splintered doorway. The horses neighed and reared up frighteningly as they charged the howling crowd. The people had nowhere to run; they were crushed, smothered and trampled. But the true horror was yet to come. The troops slashed at the defenceless bodies with their swords. Niccolò shrank down in his mother's arms, pressing himself against her. A horseman approached the pillar where his family were huddled. Niccolò watched as the man lifted his great sword, then it fell. It hit his head, his face. Unknown to the horseman, his mother was unhurt.

The battle over, French troops, led by Gaston la Foix, a handsome young man barely twenty-two years old, seized the village. They killed, raped, looted and burned all before them. Gaston died fifty-seven days later in the battle of Ravenna, his face stabbed fifteen times with a lance.

Mr Ruche trembled, remembering how he had felt fifty years ago in 1944 when he read of the SS massacre in the little church at Oradou-sur-Glane. In his studies of Grosrouvre's mathematicians, he had not expected to be confronted with images of his past.

Hundreds died in the little church at Brescia. Twelve-year-old Niccolò was carried out, his jaw shattered, two gaping wounds to his face, but he was alive. Niccolò's father was dead and his mother was too poor to pay a doctor, so she looked after him. For months, he could not speak and when, at last, he began to talk again, he stammered. His friends called him Tartaglia – the stutterer.

When he began to study Tartaglia's maths, Mr Ruche quickly discovered that if he wanted to understand it clearly he would have to work backwards a little. He began in the thirteenth century with Leonardo of Pisa, the greatest mathematician of the Middle Ages.

Leonardo was happy simply to be known as 'Bonaccio's son' – *filius Bonacci* – which, in time, became Fibonacci. It was under this name that he became famous as the author of the first great book on mathematics in Western Europe. It was called *Liber abaci*, 'Book of the Abacus'.

He learned Arabic in Algeria – something which was a great advantage to anyone studying maths, as even the Greek classics existed now only in Arabic translations. He travelled to the Middle East, which had become a required pilgrimage for every mathematician, and it was here that Fibonacci was converted – to Arabic numerals.

On his return to Europe, he championed them in his book. Christians discovered the number zero for the first time, and learned, too, about the positional number system. They learned how to factor numbers, and about prime factors and divisibility by 2, 3, etc. They also learned something about rabbits.

Fibonacci was interested in the rate at which rabbits multiplied. He wondered how many descendants a single pair of rabbits might produce in the space of a year. If their reproductive cycle begins in January, he calculated, a second couple would be born in February, which in turn would produce one new pair a month. Each succeeding couple will give birth to a couple in their second month and every month thereafter. Fibonacci noted the growing numbers of pairs like this: 1, 1, 2, 3, 5, 8, 13, 21, 34, 55, 89, 144, 233.

In a single year, Fibonacci calculated, his rabbits would produce 232 descendant pairs. Each number is the sum of the previous two.* In explaining the rabbits' reproduction like this, Fibonacci gave rise to the first *sequence* of numbers, a concept which was to have a promising future.

Stranger still, if one compared any number with the previous one, the resulting ratio got closer to a constant value, the further along the sequence one went. The constant was

$$1 + \frac{\sqrt{5}}{2} = 1.61803\ldots$$

This was the famous *golden mean*!

Mr Ruche wondered whether he should give a presentation based on what he had just learned, but decided it was a little premature. He would need to prepare the ground and rehearse with Max and Sidney before he could do so. Immersed in his studies, Mr Ruche had not given a second thought to the men from whom Sidney had been rescued. He could not know that they thought about Max all the time.

When the Short Stocky Guy received the fax sent from Tokyo by the Tall Stocky Guy, he slipped the photograph into his shirt pocket and went to the pet shop on the Quai de la Mégisserie looking for Giulietta, the girl who had phoned them some months before. He was told that it was her day off, so he went to her home.

There was no answer when he rang the doorbell, so he decided to wait in the café opposite. He ordered a beer and sat daydreaming about Tokyo. He hated the fact that he was always left behind in Paris to do the little jobs which, he thought, were beneath him. He was deep in thought when someone tapped him on the shoulder, making him spill his beer on his shirt.

* The second '1' can be thought of as being obtained by adding an imaginary initial zero to the first '1'.

'Giulietta!'

'What are you doing here?' she asked.

'Waiting for you. Something's come up.' He took the photo from his shirt pocket. Max and Sidney had been circled with a marker. 'Is this the kid who came into the shop?'

'That's him.'

'You're sure?'

'I never forget a face. I'd recognize the little brat anywhere. I could have slapped his face when he said his mother told him he shouldn't talk to strangers.'

'Don't worry. When I catch up with him, I'll slap him for you. In the warehouse at Clignancourt, he kneed me in the stomach – I was throwing up for two days. As for that parrot – *zitt*!' He twisted his hands to show he would gladly throttle it. 'See what it did to me?'

He held out the little finger of his left hand, where a savage bite was just beginning to heal. 'I'll feel a lot better when I get my hands on that bloody parrot.'

The background work on Tartaglia had taken longer than Mr Ruche expected. He was still firmly in the Middle Ages, studying a Franciscan monk named Luca Pacioli. His book, *Summa de arithmetica, geometria proportioni et proportionalità* was a masterpiece. It was published in 1494, at the height of the Renaissance. In Pacioli's work, the Arabic *chei*, the unknown, had become the Latin *cosa*. The square of the *cosa* was the *censo*, its cube was the *cubo*. A second-degree equation would be written out in words: *censo et cose egual numero*. A square and an unknown equal a number. An abridged version of the third-degree equation (without the square) would be written: *cubo et cose egual numero*. A cube and an unknown equal a number. This equation took up much of the time of the Bologna school in the sixteenth century. Also in Bologna, Sienna, Venice and Florence, Leonardo da Vinci, Raphael and Piero della Francesca were painting works which would form the core of Western art for centuries to come. Gutenberg had printed the very first book only forty years earlier. In the museum at Naples there is a portrait of Luca Pacioli, with one hand on the *Summa*: the first book about algebra ever printed.

Mr Ruche wondered what it must have felt like to be a bookseller seeing the first printed book arrive in your shop. A man who had only ever seen or held a vellum manuscript looked at a printed book for the first time. He might have thought that it was poorer for the regularity – the monotony – of the type. He imagined himself in a bookshop on the rue des Escoliers in 1480, near the Sorbonne, where the first French books were printed...

> This first book of algebra in print contained nothing that was really new, but presented, for the first time, everything that was known about algebra in fifteenth-century Europe. Much of this knowledge had come from the works of Arab mathematicians and the translations they had made of their Greek predecessors, but the work of Al-Khayyām and al-Tūsī was almost completely unknown.

The mention of al-Khayyām reminded him of the question Perrette had asked about whether he and Grosrouvre had had a third friend. 'There was someone', he thought, 'an Italian – Tavio. He worked behind the counter in the tobacconist's shop at the Sorbonne. He was a nice lad, younger than us, and good friends with Grosrouvre. For months – a year, perhaps – we were inseparable. That was before war was declared. Grosrouvre and I signed up. We never saw him again. It was a short-lived trio.' Mr Ruche had thought a lot about his past these last days, and he could think of no other trio. He couldn't imagine Tavio being interested in mathematics. No, Perrette had to be wrong. Mr Ruche went back to his work.

> Al-Khwārizmī was the most famous mathematician in Europe in the Middle Ages. His books had been translated since the twelfth century. *Algorithmi de numero indorum*, his book on the Indian system of calculation, had become the bible of mathematics, so much so that calculation became known as 'algorism', from which the word 'algorithm' derives.
>
> Roman numerals were ill-suited to calculating. Even the simplest task had to be performed on an abacus, which was often simply rows of columns on a board on which small tiles were placed.
>
> The introduction of this new system of calculation started a revolution. Those who were in favour of the new system – 'algorists' – fought a pitched battle with the 'abacists' – professional calculators trying to protect their privileged status.

'Performing an operation' – the simplest of mathematical tasks, which involved writing down a calculation and working it out – would have been impossible for most people at the time. In the twelfth and thirteenth centuries, the ability to calculate and to multiply opened doors to the highest levels of society.

The great change came from no longer using objects – pebbles or tiles – to count, but words. People began to count using the *names* of the numbers themselves. This radically changed the nature of calculation.

It had never occurred to Mr Ruche that there had ever been any other form of calculation. But there was a greater shock to the system: the arrival of *zero*!

The road from the *idea* of zero to the number we recognize today was a long one. Originally, numbers were represented by one of the nine numerals and placed in columns to indicate units, tens, hundreds, etc.

As the owner of A Thousand and One Pages, Mr Ruche had a perfect example: one thousand and one. He scribbled on a scrap of paper:

If the bars separating the digits were removed, the number collapsed:

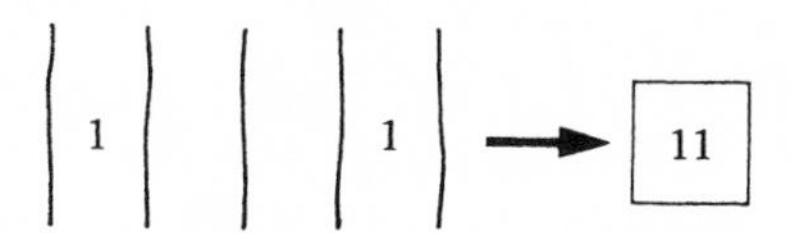

and the number 'one thousand and one' became 'eleven'. Then, somebody invented a symbol for an empty column: a small circle. Mr Ruche went back to his sketch:

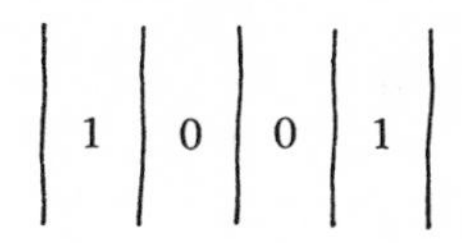

It was now possible to represent something that wasn't there. Nothing could be treated as something. This symbol, rather than being something apart – like a punctuation mark – became a numeral like the other nine. Now, if Mr Ruche took away the lines indicating the columns, the zeros prevented the number from collapsing:

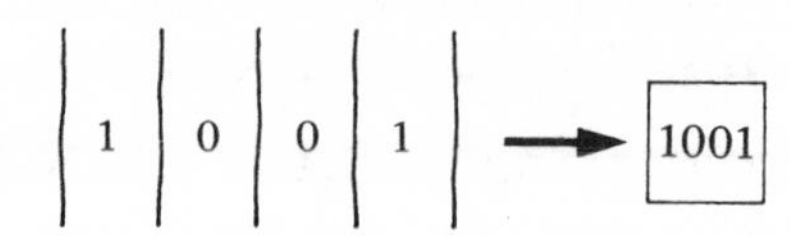

The number zero existed in Babylon in 300 BC, where scribes wrote it as a double arrow. Later, a small circle was used in Greek maths to represent zero, possibly originating as the initial letter of the Greek word *ouden*, meaning 'nothing'.

It was not until the sixth century AD that the 'true' zero was invented, which was not just a symbol but a number, a number in the sense that it can be a part of an operation. This was the invention of the number 'nothing', the great invention of the Indians. *Sunya*, or zero, was defined as the result of subtracting a number from itself:

$$0 = n - n$$

In philosophical terms, zero was the difference between the thing and itself.

It had no effect on addition: $n + 0 = n$.

But a staggering effect on multiplication: $n \times 0 = 0$.

It was impossible to divide using zero: $\cancel{\frac{n}{0}}$.

But devastating when used as a power: $a^0 = 1$, if $a \neq 0$.

These were the actions of this new number.

In answering the question 'how many are there?', zero changes the negative 'there are not any' into the positive 'there are zero'. This completely changed the world of numbers – 'nothing' was now a number.

Between Fibonacci and Pacioli, something important had happened. In 1453, the armies of the Sultan Muhammad II seized Constantinople. The fall of the 'Middle City' – so called because it stood between Rome and Baghdad, between Christianity and Islam – was to have unexpected consequences. Hundreds of Byzantine scholars and translators fled the city, taking with them Greek manuscripts which were to profoundly influence the course of Western mathematics.

This brought Mr Ruche full circle back to Tartaglia, for this influx of knowledge was crucial in his education. The Turk had become the enemy. In fact, in a book of mathematical games – a novelty at the time – Tartaglia posed the following problem:

> A boat carrying 15 Turks and 15 Christians is caught in a storm. To stop the boat from sinking, the captain orders that half of the passengers be thrown overboard. To find out who is to be thrown overboard, the passengers are placed in a circle and, starting at a particular spot, every ninth passenger will be thrown overboard. How should the passengers be placed so that only Turks will be thrown into the sea? To solve the problem, the captain had to use algebra.

Mr Ruche smiled, and continued his note-taking.

> Tartaglia was interested in solving third-degree equations using *radicals*. This involved looking for formulas which used only the four basic operations – addition, subtraction, multiplication and division, and radicals – square roots, cubes, etc. Al-Khayyām had succeeded in solving third-degree equations only by using geometric constructions. He had hoped that later mathematicians would solve equations using 'pure calculation'. This was exactly what Tartaglia was trying to achieve.
>
> The first person to find a mathematical solution to third-degree equations, however, was Scipione del Ferro. Rather than publish his method, he kept it secret.

Mr Ruche opened *Questiti e Invenzioni diverse*, where Tartaglia explained the sad story of his attempt to solve the third-degree equation. Leafing through the book, Mr Ruche noticed little crosses in the margins. Grosrouvre had marked this passage, perhaps to indicate that he was not the only one to keep his results secret. Scipione explained his method to his son-in-law, Annibal de la Nave. Grosrouvre had not told anyone his methods – not even his oldest friend.

Annibal explained the method to one of his friends, Anton Maria Del Fiore, who kept the secret until 1526, when Scipione died. After that, he tried to use the secret to his advantage – issuing challenges to other mathematicians.

Tartaglia took up the challenge, and an algebraic duel began between the two men. Each submitted a list of thirty problems and a sum of money to a lawyer. Whichever of them solved the most problems in the forty days that followed would keep the money. We know Del Fiore's thirty problems. 'Find a number, which, added to its cube root equals 6', or 'The winnings of two men amount to 100 ducats. The winnings of the first equal the cube root of the winnings of the second', or 'A Jew lends a sum of money on condition that, after a year, he be repaid with the cube root of the capital as interest. At the end of the year, he receives 800 ducats, representing capital and interest. How much did the Jew lend?'

All the problems set by Del Fiore involved third-degree equations, to which he held Scipione's secret. Tartaglia, however, solved them in a matter of days. Del Fiore did not succeed in solving any of the problems set by Tartaglia, but he contested the result of the challenge. Tartaglia was still declared the winner. Everyone waited for him to publish the method by which he had won so easily. Tartaglia did not publish, however, though he insisted that he had no wish to keep his method secret and that he was preparing a book on the subject, which he would publish shortly afterwards.

It was at this point that a Milanese doctor entered the story. Doctor and mathematician, Girolamo Cardano, also known as Jerome Cardan, was born in 1501 in Pavia. Cardano's life was a story of bad luck: as a baby, he caught smallpox, at eight, dysentery, and at nine, he fell down a flight of stairs, splitting his forehead open. At eighteen, he was struck down by the plague. Twice, he almost drowned.

Despite all this, he grew up to be an intelligent man. By the time he was twenty, he was teaching Euclid at the University of Pavia. Like his father, Girolamo was a doctor and a mathematician.

Twice in his career, he burned some of his writings. The first time he burned nine books, the second time, one hundred and twenty-four. There remained, however, fifty books in print and at least as many manuscripts.

Grosrouvre had been much more radical than Cardano, thought

Mr Ruche. He had burned all his papers, his notebooks, his jottings – his whole life's work. Mr Ruche began to appreciate the state his old friend must have been in as he wrote his second letter. He imagined Grosrouvre writing and, from time to time, looking over at his manuscripts piled in the middle of the room. His letter truly was his last will.

When Cardano heard of Tartaglia's magisterial success in the algebraic duel, he wrote to him. For several years he tried to convince Tartaglia to divulge the secret of his methods, but Tartaglia refused. Cardano tried trickery, prayer, even threats. He was furious at being refused and wrote to Tartaglia accusing him of being 'bloated with self-importance, like someone who thinks himself at the top of a mountain when he is still in the valley'.

In time, however, Cardano succeeded in becoming Tartaglia's friend and was permitted to see some of the problems he had set Del Fiore. Others, Tartaglia kept secret.

Although Cardano had worn Tartaglia down a little, the man still refused to tell his secrets. But Cardano had a trump card: he was a doctor. For Tartaglia, who had suffered so much as a child and who had never had the attentions of a doctor, it was enough to break down any barrier.

In 1537, Tartaglia published *Nova scientia*. People bought it, excited to discover the formulas that had allowed him to solve the equations. They were disappointed to discover that there was not a word about them. The book was about manufacturing explosives. This was indeed the beginning of a new science: ballistics.

Tartaglia had worked on the trajectory of cannon balls, fascinated to discover the relationship between the distance travelled by a projectile and the angle at which it had been fired. Tartaglia came up with two results:

The trajectory of a cannon ball is never a straight line, but the faster it travels, the less curved is the trajectory.

The maximum distance that can be achieved is when the cannon ball is fired at an angle of 45°.

Tartaglia's formulas, however, had still not been published, and Cardano became more insistent. Cardano made a promise: 'If you will explain your proofs to me, not only will I swear never to publish them, but I will write them in code such that, after my death, no man will be able to understand them.'

Grosrouvre had put a cross beside this passage. Mr Ruche paused. Had Grosrouvre written down his proofs in code, he wondered? If he had, his 'loyal friend' might only have a copy of the coded text. If this proved to be true, they would have to identify the friend and decode the proofs. He went back to the book.

One day in March 1539, Tartaglia finally agreed. Cardano could feel his heart beating faster. He sat and listened as his friend, still stammering slightly, began:

'You wish to solve the equation "a cube and something equals a given number". You must find two numbers whose difference is equal to the given number and whose product is the cube of one-third of the given number. The solution, then, is the difference between the cube roots of the two numbers.'

Despite the explanation, Cardano was unable to solve the equations. He said as much to Tartaglia, implying that he did not believe that Tartaglia had solved the equation. Tartaglia explained that the error was Cardano's own: he had misinterpreted Tartaglia's explanation. Cardano had been using 'one-third of the cube' instead of 'the cube of one-third'.

Shortly after this, Cardano published his *Ars magna*, 'The Great Art'. Tartaglia was eager to read his friend's book, only to find there his solution to third-degree equations described in detail. Cardano had betrayed him. Tartaglia's response was simple: '*Quello que tu non voi che si sappia, ne dire ad alcuno*' – 'If you wish to keep a secret, tell it to no one.'

In the margin, Grosrouvre had put two small crosses. Grosrouvre had followed Tartaglia's advice. The *Questiti* ended without a word about the great treatise that Tartaglia was to publish later. Mr Ruche was disappointed.

The first of the six volumes of the *General trattato di numeri et misure* was published eleven years later. The first four parts were published in 1556. The fifth volume was prepared for printing, but before it was published Tartaglia died. The sixth volume, which was to deal with

> Tartaglia's solution to third-degree equations, was never published and no trace of it was ever found.

Mr Ruche was shocked. If Cardano had not betrayed Tartaglia and published the formulas, they would have been buried with him. Tartaglia's formulas are among the most important contributions to mathematics, but they are famous as *Cardano's* formulas.

Mr Ruche quickly turned the pages, eager to see the formulas, but when he did, he was disappointed. He had expected to see formulas with *x* and *y*, *a* and *b*, and the whole host of mathematical symbols; instead, he found something which looked like prose. Not a single '=', only 'aeq' standing for *aequalis*, and 'P' standing for 'plus'. He returned to his note-taking:

> Cardano, in his *Ars magna*, went further than Tartaglia. Not only had he published his friend's formulas, which were valid only for specific equations, he gave others. He was the first to present a complete solution to third-degree equations using radicals.
>
> There was something else remarkable in the *Ars magna*: *fourth*-degree equations could also be solved using radicals. This discovery was due not to Tartaglia, nor to Cardano, despite their best efforts, but to Ludovico Ferrari. Cardano had hired Ludovico as an errand-boy when he was fifteen. Because of the interest Ludovico showed in Cardano's work, the master allowed him to attend his lessons. Ludovico studied so well that he surpassed his master, a man who loved him dearly.
>
> Third- and fourth-degree equations had now been successfully solved – would fifth-degree equations and beyond prove soluble using radicals?

Several days later, when he had absorbed his notes, Mr Ruche reported his findings to the others so that, together, they could analyse the information for links to Grosrouvre's story. The most important question was, as always, 'How does this advance the investigation?'

He concluded the session with the title of Fibonacci's book, '*A Flower of Solutions to Certain Problems Concerning Numbers and Geometry*'. 'The name', he explained, 'came from Fibonacci's statement that many problems, "though thorny, flower until they are solved, rise up like plants and put forth many flowers, so from these questions, hundreds of others will follow." '

'This is getting us nowhere!' Jonathan exclaimed suddenly. 'We already had three problems to solve: squaring the circle, doubling the cube and trisecting an angle, and now we're trying to solve algebraic equations using radicals when we're still no closer to finding the answers to the first three problems, and we don't even know if they *can* be solved. We can't just go on adding more problems – we'll never get anywhere.'

Mr Ruche tried to keep a straight face as Jonathan went on: 'Mr Ruche, young people today...The youth of today need...'

Lea held her nose to stop herself from laughing.

'They need answers, they need something solid, something to live by. We need to solve the problems we have already. And another thing – what about the language we're using in the equations? It's all very well to say the cube of this and the *cose* of that, but I don't understand a word of it. We should use the same terms we learn at school.'

'You have to run before you can walk', said Mr Ruche. 'Tartaglia was writing in the sixteenth century, not the twentieth. History is about how things came to be as they are today.'

'Surely history is also about what might have been?' asked Lea, cleverly.

'Yes, yes it is. You're right', said Mr Ruche. 'History is also about the possibilities which never came to pass, about paths that were never taken.'

CHAPTER 16

Less is More

Signs and symbols; imaginary numbers & solving Cardano's equation

Grosrouvre's copy of *A History of Symbols and Notation in Mathematics* and some of Cardano's books were strewn over Lea's bed. The twins had decided to show Mr Ruche what they were made of: they had decided to work out Cardano's formulas. The way Mr Ruche had presented them made them impossible to understand. The twins were about to give them a 'makeover'. Jonathan read aloud:

'In a poorly furnished office lit by a single candle, Robert Recorde sat poring over a page dark with scribbled numbers and shapes. Quill in hand, he thought for a long time, then, dipping the quill into the inkpot decisively, he drew a short horizontal line. He lifted the quill and carefully drew a second, parallel line above the first.

'He set down the quill, picked up the paper and held it at arm's length. He studied the symbol he had just drawn. In front of him was the most famous symbol in the history of mathematics, the equals sign. Two short parallel lines separated by a thin cushion of air:

$$=$$

The year was 1557, and for some time mathematicians had been discussing the need for a symbol to replace the word *aequalis* – "equal" – in written equations. What could they use to symbolize this familiar and yet complex concept? Later, Recorde was asked why he had chosen this particular symbol. "I chose parallel lines, because they are twins and nothing is more alike than a pair of twins."'

Jonathan and Lea exchanged a look. What the Liard twins saw when they looked at one another were their differences: the tiny differences which simply heightened their similarity. They were not the same in the sense that two printed books are the same, but in the sense that two manuscripts written by the same scribe are similar.

'When did the plus and minus signs get invented?' asked Lea.

'Hang on a minute, we haven't finished with Recorde yet', replied Jonathan. 'Listen: shortly after inventing the equals sign, Recorde was thrown into prison for unpaid debts. He died there some months later.'

'You're not serious?' Lea laughed. 'You mean the guy who invents the equals sign dies in prison because what he spends isn't equal to what he earns?'

'One of the little lines must have been longer than the other.'

'More like "All mathematicians are created equal, but this one was more equal than others".'

They went back to the signs and symbols. Jonathan explained that the + and – signs first appeared in a book about arithmetic in business.

'It was in 1489 and the author, Johan Widman, used them to mark his crates of merchandise. The crates were called *largels* and, once filled, each was supposed to weigh 4 *centner*. When the crates did not come to the exact weight – if one was five pounds light, for example – he would mark it "4c – 5l". If, on the other hand, the crate was five pounds heavier, he put a cross through the symbol, making "4c + 5l". From the wooden crates, the symbols made their way into his accounting books, and from the world of business they found their way into the world of algebra.'

Lea lay on the bed, eyes closed, listening. When Jonathan had finished, she said, 'So minus came before plus – in fact, plus was simply a minus crossed out.'

'Less is more', said Jonathan philosophically, showing Lea the hieroglyphics that the ancient Egyptians had used for addition and subtraction.

Jonathan went on listing symbols. The multiplication sign × was invented by William Oughtred, in 1631; the signs for 'more than', >, and 'less than', <, had been invented by Thomas Harriot. The √

used for square roots was devised by a German called Christoff Rudolff in 1525. Originally, there were three for a cube root, $\sqrt{}\sqrt{}\sqrt{}$, four for a fourth root...

'What about infinity?'

'An infinite root?'

'No, the symbol for infinity.'

Jonathan leafed through the book and found the right page. This was another symbol invented by an Englishman, John Wallis, who was not only a mathematician but a doctor too. The symbol was a horizontal eight: ∞.

Jonathan moved onto *exponentials*, telling Lea in minute detail about Nicolas Chuquet's *Triparty en la science des nombres* – the earliest known text about algebra in French.

'Guess what he did for a living?'

'Doctor?'

'That makes four of them now. It's all very well Mr Ruche saying mathematicians are poets, but they're more likely to be doctors. Anyway, when Chuquet wanted to write "2 to the power of 4", he simply rubbed out "to the power of" and moved the 4 up a bit: 2^4. If the power was part of a denominator, he simply made the denominator the numerator and added a minus sign. Clever!

$$\frac{1}{2^4} = 2^{-4}$$

'He was using negative exponentials centuries before mathematicians even used negative numbers. Listen: "If a man would subtract minus 4 from 10, the result is 14 as though a man who has nothing owes another 4." A negative number is like having nothing but still owing something.

'Negative numbers mean that you can add things together and end up with less than you started with', Jonathan continued. 'If you've got "minus three" things, it means you haven't got anything and you still owe me three.'

'Which is pretty much what happened to Recorde. Negative numbers lead to jail and directly to jail. If zero is nothing, then minus numbers are less than nothing.'

'He was ahead of his time, this Chuquet guy, but guess what? He never got to publish his *Triparty*. Nobody even read it at the time, so it didn't have any influence on anything.'

'The more we go on, the more it looks like Grosrouvre wasn't the only one who didn't publish his research.' Lea was thinking aloud. 'What about the letters?'

'Oh, let's not go into all that again. We're always talking about Grosrouvre's letters.'

'How on earth did I end up with a twin like you? I was talking about letters used in equations.'

Jonathan leafed through the book again, then after a minute or two continued: 'OK, the hero in this case is someone called François Viète, known as "the Man of Letters". Before him, people sometimes replaced quantities with letters, but only unknown quantities. Viète used letters everywhere. He only used capital letters – vowels for the unknown numbers and consonants for the known ones.

'France, under the reign of Henry IV, was in the middle of a religious war. The Duc de Guise had been assassinated. One day, some of the king's men intercepted letters from the Spaniards. The letters were in a code which used more than 500 characters. They were impossible to decipher. Henry sent the letters to Viète.

'More letters were intercepted, and the Spaniards changed the code several times, but Viète established a system which allowed him to track changes in the code. Convinced that no one could decipher the code without using magic, the Spanish denounced Viète to the Inquisition. He was brought before the Holy Inquisition in Rome as a sorcerer. Coincidentally, this happened at the same time as Cardano was imprisoned by the Inquisition. They seemed to have a taste for mathematicians.

'Right. Skip a couple of decades', Jonathan went on, 'and we come to René Descartes. He replaced capitals with lower-case letters and decided that the early letters of the alphabet, a, b, c, ..., should represent the known quantities, and the last, x, y and z, should represent the unknowns.

'OK, now that we're researched the history of the symbols needed for Cardano's equations, we can write them out again.' Jonathan hastily scribbled out the equations as he spoke. 'We'll need to simplify them – move everything from the right-hand side of the equation to the left so that the only thing on the right is zero.'

He rewrote the equations neatly. 'And what do we come up with?' said Jonathan with a flourish, '*ax* squared plus *bx* plus *c* equals zero! Lea? Are you listening? Right, now let's get to work', said Jonathan grabbing Cardano's book, but Lea was already asleep. Jonathan worked alone through the night translating Cardano's interminable formulas 'borrowed from Tartaglia' into language a schoolchild would recognize. It was dawn by the time he had finished. He put away the papers, opened his skylight and swept the snow from the outside. He slept for an hour or so, and slipped the piece of paper under Mr Ruche's garage door on his way to school the next morning.

On the far side of Paris, the Short Stocky Guy was tearing open a letter with a Tokyo postmark. In it was a translation of the Japanese magazine caption that had appeared beneath the photo of the parrot:

> A venerable French mathematician measures the Louvre Pyramid, designed by architect Ieo Ming Pei, using the ancient shadow-method of the Greek mathematician Thales.

'What the hell do I care about what the caption says?' All the same, the Short Stocky Guy went to the Louvre. But, despite bribing a couple of the doormen, he couldn't find out anything about the 'venerable French mathematician' or about Thales. He made half a dozen photocopies of the photo and posted a couple of men on either end of the Quai de la Mégisserie for a day or two in case the kid came back.

At lunch he met Giulietta at a café for a beer. As they were talking, an idea came to him. Kids go to school, don't they? The Short Stocky Guy wondered how old the kid was. Giulietta was sure that he was about eleven or twelve. He would probably be in his first or second year in secondary school. He thought for a moment and then said, 'Do you know how many secondary schools there are in Paris? There's no way I could check out every kid in every class.'

'Who said the kid lived in Paris?' Giulietta enjoyed winding him up. 'He could live in the suburbs. Kids come into town to the market all the time.'

The Short Stocky Guy's face fell.

'But', she added, 'as I said before, there was something...something odd about him; about the way he looked at you when you talked to him. Looked really carefully as if...'

'Maybe he fancied you', said the Short Stocky Guy with a smirk. 'He wouldn't be the first one.'

'I don't know', she said, 'but he had a strange effect on me.'

'He's a bit young for a toyboy.'

'God, you're stupid!' She turned on her heel and walked out.

The Short Stocky Guy sat, sipping his beer, and thought for a long time. Suddenly, he came up with another idea. He had a photo of the kid, the kid had to be in school, and every year schools took class photos. He could get a list of photographers from the phone directory and try to find the kid that way. He tapped his forehead. 'Mom didn't raise stupid children.'

The list was easy enough, but when he tried calling on the photographers, they became suspicious. The photos were private, they argued, and in any case the children were minors. The Short Stocky Guy was prepared for this. He was, he said, working freelance for a Japanese bird-lovers' magazine, and he produced the copy of the photograph and pointed to the parrot on the boy's shoulder. The bird – and the boy – had won a prize, he said, and the magazine was trying to get in touch with him. It was a tidy sum, he added.

'Of course, there will be a reward for whoever can help me get my hands on...can help me find them.' He left copies of the photo with each of the photographers. Now all he had to do was wait.

Later that morning, Mr Ruche picked up the piece of paper that Jonathan had slipped under his door. He opened it and was surprised to find the following:

This is how the Egyptians noted mathematical operations:

Addition
Legs walking in the same direction as the text

Subtraction
Legs walking in the opposite direction

Mr Ruche thought for a moment, and then remembered that Egyptian texts ran from right to left. He smiled, threw a rug over his legs (which didn't walk one way or the other) and went to his shoe rack. As he took down a pair of boots, he re-read the quote from Plato above the rack: 'It is impossible to understand the science of shoes until one understands what science is.'

The rest of Jon-and-Lea's message was a little more down to earth: 'We're still running before we can walk, as you put it: this is what Cardano's formula should look like.'

Mr Ruche looked at the equation. This was exactly the sort of thing which had scared him stiff when he was a student:

$$\sqrt[3]{\left[-q/2+\sqrt{(q/2)^2+(p/3)^3}\right]}+\sqrt[3]{\left[-q/2-\sqrt{(q/2)^2+(p/3)^3}\right]}$$

The twins had succeeded in writing out the equation in modern notation. It took Mr Ruche some time just to understand it. The formula produced more solutions than expected, and some of them were blind alleys. Practically speaking, it couldn't be applied. He went back to his books:

> One day, a friend of Tartaglia told him that he found it difficult to believe that a third-degree equation could have two solutions and sometimes more. 'True, it is difficult to believe', said Tartaglia, 'and if experience did not bear it out, I should scarcely believe it myself.'

So, thought Mr Ruche, a third-degree equation could have more than one solution. But how many – two, three, more? Everything, he discovered, depended on negative numbers.

If $(q/2)^2 + (p/3)^3$ was negative, the equation would be impossible to solve because it is impossible to extract the square root of a negative number. Mr Ruche remembered why. The square of a number is always positive, regardless of whether the number is positive or negative: + and +, or – and –, always made +.

The square root of the number a is $\sqrt{a}$. It is a number which, when squared, equals a: $(\sqrt{a})^2 = a$. What if a is a negative number? Then the square would be negative, which was impossible:

> A negative number cannot have a negative square root.

Mr Ruche went back to his original notes on Cardano:

> When the equation produced a negative result, it was impossible to solve and there are no possible roots. But, reading Archimedes' *On the Sphere and the Cylinder* – perhaps even in Tartaglia's own translation – Cardano realized that in this case he had shown that there were three roots.
>
> Cardano checked his work. (1) The formula was correct. (2) There was a particular case in which it could not be applied, and that directly contradicted Archimedes' findings. (3) The reason this was so is that it is impossible to find the square root of a negative number.
>
> To Cardano, the solution was clear. He advised his readers, 'forget your mental tortures and simply insert these quantities into your equations.' He introduced expressions like '$\sqrt{-1}$', and it worked. Mathematicians had spent hundreds of years trying to work out $\sqrt{2}$. Now they were faced with $\sqrt{-1}$!
>
> The first to work on it was Rafael Bombelli. He had fewer scruples than Cardano about using UMOs – unidentified mathematical objects. He decided to work with negative roots using the same rules as with 'normal' numbers. His book, *Algebra*, in which he presented these new ideas, immediately eclipsed the works of Cardano and Tartaglia. Bombelli, however, did not have time to profit from his fame as he died in the year it was published.

In passing, Mr Ruche noted, Bombelli pointed out that the trisection of an angle was, in effect, a third-degree equation. This was new, but it did not solve the problem of performing it with a ruler and compass. The information might, however, be of great importance: a problem which until then had been purely geometric had jumped the fence to become an algebraic problem.

There was another thing. Bombelli had invented something very useful which Jon-and-Lea had not included in their list of symbols: brackets. Brackets were the great unsung heroes of mathematics.

Brackets work in pairs – one opens, the other closes. They are essential in writing down equations that are unambiguous. Mr Ruche scribbled an example using two divisions combined. 2 divided by 3 divided by 5 equals how much? If he wrote it as 2/3/5, it was meaningless. It could mean 2/3 divided by 5, or 2 divided by 3/5. How could he be sure? Without brackets, he couldn't.

With brackets, you could write the equation as (2/3)/5, which equals 0.13333333333..., or the brackets can go at the end: $2/(3/5)$, which equals 3.33333333333..., not at all the same thing.

This was the mistake Cardano had made with Tartaglia's work. Cardano thought he had meant the 'third of the cube'; in fact, it was the 'third cubed'. With brackets, he would never have made the mistake: $(p^3)/3$ could not be confused with $(p/3)^3$.

Mr Ruche went back to his notes:

> Before Bombelli, there had been the pair +1, −1, *più* and *meno*. Bombelli added another pair: *più di meno*, $+\sqrt{-1}$, and *meno di più*, $-\sqrt{-1}$. Having set down the new scheme, he made up a rhyme to help popularize it:
>
> *Più di meno via più di meno fa meno.*
> *Più di meno via meno di meno fa più.*
> *Meno di meno via più di meno fa più.*
> *Meno di meno via meno di meno fa meno.*
>
> Which means
>
> $$\sqrt{-1} \times \sqrt{-1} = -1$$
> $$\sqrt{-1} \times (-\sqrt{-1}) = +1$$
> $$(-\sqrt{-1}) \times \sqrt{-1} = +1$$
> $$(-\sqrt{-1}) \times (-\sqrt{-1}) = -1$$
>
> It was now possible to calculate using completely new objects, though no one seemed keen to define them since they seemed fictitious. They were simple intermediaries used for calculating which disappeared afterwards without trace. Descartes called them *imaginary numbers*. The numbers that had been used until then – positive, negative, rational and irrational – were referred to as *real numbers*. Leonhard Euler, in 1777, replaced $\sqrt{-1}$ with the symbol we are familiar with today: $\sqrt{-1} = i$, 'i' for 'imaginary'. Later still, once their 'reality' was taken for granted, the German mathematician Carl Friedrich Gauss would introduce the term *complex number*.

Mr Ruche scratched his head. He was certain that Euler was one of the mathematicians Grosrouvre had mentioned in his letter. He checked, and found that Euler came on Grosrouvre's list just after Fermat, who followed on from Tartaglia.

He thought for a long time about these strange mathematical objects which moved from impossible to imaginary and from imaginary to complex. What did a complex number look like? A complex number had a real part and an imaginary part. For example, from the pair (2, 3), one could make the complex number

$$2 + 3i$$

With the pair (2, 0), one generates the number $2 + 0i$, which equals 2. This implied that a real number was simply a particular form of a complex number. In fact, real numbers had simply been plunged into a larger set of numbers. The universe of possible numbers had been expanded to make the impossible possible.

One thing still bothered Mr Ruche. Was it possible to find the square root of a negative number or not? Well, yes *and* no! No, it was not possible to extract the square root of a negative number in the world of real numbers. But yes, it *was* possible to find the square root of a negative number using complex numbers.

Lastly, what exactly was 'i'? It was, according to mathematicians, 'an imaginary root of a negative number'. Since he had begun his studies, Mr Ruche realized, he had come up against this concept more than once – the almost philosophical idea of existence and impossibility.

He tried to sum up the problem:

> At certain times in history, some mathematicians faced with problems they could not solve have secretly used illicit methods. They were thus no longer subject to the same laws, and could use methods which, though disturbing, were effective. The mathematicians who followed them expanded the world of mathematics to accommodate these new mathematical ideas.
>
> There was no such thing as 'pure notation'. To write the 'impossible' is to legitimize it, to dare to question whether it might exist. Such new ideas can always be accommodated as long as they do not call into question those which are already there, nor invalidate results which have already been established.
>
> Mathematical revolutions do not destroy the old world, but construct new worlds which encompass those they replace.

When Mr Ruche explained to Jon-and-Lea what he had discovered, they were quick to react.

'But that's exactly the opposite of what you said before', said Jonathan, 'about the ruler and the compass, when you said "Without ruler and compass, nothing may be constructed." '

'And now with imaginary numbers, you don't seem to care how you get your result as long as you get one. "The end justifies the means." You just ignore the means and...' Lea didn't finish her sentence.

'Well, once you've got a result, the means don't matter', said Jonathan. 'The only thing that matters is that it works.'

Mr Ruche spun round in his wheelchair. 'What if it doesn't work?'

Lea smiled at him: 'If it doesn't work, you cheat!'

Sidney landed on Lea's shoulder – something he had never done with anyone besides Max. Lea was embarrassed.

The next day, Jon-and-Lea took things in hand. Since Mr Ruche hadn't arranged a session on the subject, they decided to do one themselves. They knew they could count on Mr Ruche, Perrette and Max to come, and they also invited Albert and Mr Habibi. Sidney was already involved.

Sitting on his highest perch, Sidney began by doing a slow somersault. Once he was upside down, he announced, 'The Imaginary Drama'. He jumped down to the next perch, and swung himself upright. Thrusting out his long neck, he ruffled the red feathers at his wingtips and said: 'A play in i acts'.

Over the music of the Volga boatmen, Jon-and-Lea came forward, chanting in time to the misery of the galley slaves rowing in the depths of the ship. When the music stopped, they picked up a Persian theme and, inspired by al-Khayyām, recited a poem:

Imaginary numbers.
What strange mathematic wonders,
Condemned at first as outlaws,
They could not steal your thunder.

They solved algebraic equations
That had long been occasions
for mathematicians' squabbles
And magic incantations.

They proved a great solution
In the numbers revolution,
And earned their rightful presence
In this short contribution.

They might well be neglected,
They're imaginary and rejected.
But they work like other numbers
And deserve to be respected!

Perrette listened in silence to the strange story of imaginary numbers, and Mr Ruche was very impressed by Jon-and-Lea's work. After al-Khayyām's quatrains, Tartaglia's triplets and Bombelli's rhymes, the Rainforest Library was beginning to foster its own poetry.

CHAPTER 17

Fame!

Niels Abel and Évariste Galois; the Paris Institute & the long-awaited proof

The session had concluded with a discussion as to whether fifth-degree equations could be solved using radicals. They decided that one of them should work to find out. The fact that they had not yet solved any of the three problems of antiquity loomed large in the discussion. After all, they couldn't keep on failing to solve the problems they posed themselves.

Mr Ruche was elected to set to work, and the following morning he settled himself at his desk in the Rainforest Library, took out his glass fountain pen and his big notebook, and began to write:

> Solving equations using radicals is concerned only with *algebraic* equations which contain only polynomials. For example:
>
> $$2x^2 + 3x + 1 = 0$$
>
> is a second-degree algebraic equation;
>
> $$\sin x + 1 = 0$$
>
> is not. The most general form for an algebraic equation is
>
> $$a_n x^n + a_{n-1} x^{n-1} + \cdots + a_2 x^2 + a_1 x + a_0 = 0$$
>
> Here, n is the *degree* of the equation, and the coefficients a_i are numbers.
>
> For the first algebrists, things were simple. Equations were either soluble or insoluble: either they had a root or they had none. Cardano, Bombelli and others forced them to admit that things were more complex than that. And, as a result, more interesting.
>
> The first question to be posed now was how many roots could an

equation have? Before trying to determine what they were, they decided it would be useful to know how many there were. Could a second-degree equation have three solutions? A fourth-degree equation no possible roots? Was there any way one could be sure?

In his *Invention nouvelle en l'algèbre* ('New Developments in Algebra'), published in 1629, Albert Girard hypothesized that an *n*th-degree equation had *n* possible roots, if one took into account imaginary roots and counted each root each time it appeared.

Jean d'Alembert, who worked on the *Encyclopédie* (the first great modern compendium of knowledge, published in 1751–2), was the first to attempt a proof in 1746, followed by Euler in 1749. Two Frenchmen followed suit, Joseph Louis Lagrange and Pierre-Simon Laplace. In the end, it was the 'prince of mathematicians', Carl Friedrich Gauss, who presented the first complete proof. Not content to present one, he presented three others – proof if proof were needed of the difference between stating a theorem and proving it.

Now it was certain that an equation of degree *n* had not only multiple roots, but it had precisely *n* solutions: this was the *fundamental theorem of algebra*. It was an extraordinary theorem. It was simple and clear: a third-degree equation had three roots, a second-degree equation had two.

Mr Ruche thought for a moment, and remembered something from his schooldays three-quarters of a century before: 'For any second-degree equation: if the discriminant is negative, there are no roots; if it equals zero, it has a double root; if the discriminant is positive, there are two roots.' Now it seemed that all second-degree equations had two solutions, but he knew he had remembered the phrase correctly.

He could ignore the problem and continue studying, but his left brain refused to accept this blatant insult to his powers of logic. In the end, he discovered the solution: both statements were correct, but they related to different numerical universes. The phrase he remembered from school related only to real numbers, while the theorem related to complex numbers – there was no contradiction.

'There was an old paradox', thought Mr Ruche. ' "Where should one look to find what one is seeking?" My mother always said that. It's been a long time since I've thought of her. My mother said that if they sent me to sea, I wouldn't find the water. Well, in algebraic equations, looking for solutions in the world of complex numbers is like looking for water in the sea – it's always there.'

It was then that Mr Ruche realized quite how important the world of complex numbers really was. It had the sheer force of numbers – enough to provide solutions for every algebraic equation.

Mr Ruche went back to his quest to discover whether fifth-degree equations could be solved using radicals. Poring over Grosrouvre's index cards, he discovered that the problem had lasted three centuries, each mathematician passing it on to the next – del Ferro, Tartaglia, Cardano, Ferrari, Bombelli, Tschirnhaus, Euler, Vandermonde, Lagrange, Ruffini and, now, Niels Henrik Abel.

'Copenhagen, in the year $\sqrt[3]{6,064,321,219}$ (careful with the decimal part).'

Bernt Holmboe smiled when he read the first line of the letter, and knew at once that Niels Henrik Abel had written it. Excited by the riddle in the date, he started calculating. A cube root was never an easy thing to work out, but as a maths teacher he knew a lot about using logarithms. The answer was 1,823.590827 years. To work out 0.590827 of a year, multiply by 365 days = 216 days. So, it was the 216th day of the year 1823. He checked his calendar. The letter had been written in Copenhagen on 4 August 1823.

Niels was an ex-pupil of Holmboe, who at the end of Niels's first year had written on his school report, 'To his natural genius, he brings an insatiable appetite for mathematics. If he lives, he will become the greatest mathematician in the world.' Why he had added 'If he lives', he didn't know.

At twenty-one, Niels was unquestionably the best mathematician in Norway, having easily assimilated Euler's entire work. As soon as he knew enough about mathematics, he became passionate about the solution to fifth-degree equations. The old debate as to whether they could be solved using radicals had recently resurfaced. Euler, Abel's hero, had attempted a proof and failed.

Niels found a formula for solving fifth-degree equations which no mathematician who studied it could find fault with. In time, he discovered that the formula did not work in all cases. For it to be valid, it *had* to work in all cases.

At this point Niels changed his mind completely. If the formula had not been found, he decided, it was because it did not exist. He worked hard, studying the works of Lagrange, who had explored this field most thoroughly. With this background and considerable work, Abel spent six months honing his proof. Shortly before Christmas, it was finished. The result was extraordinary – written in a single simple sentence: 'Algebraic equations of the fifth degree cannot be solved using radicals.'

Abel wrote 'Some Thoughts on Algebraic Equations, in Which it is Proved Impossible to Solve Fifth-degree Equations'. He immediately sent his tract to the greatest European mathematician, Gauss, who filed it away without even bothering to read it. He wrote a second article, this time on the subject of *integration*, which he sent, with a dossier, in the hope of getting a bursary to attend university. He was awarded the bursary, but the only copy of the article disappeared and was never found.

Still, he tried everything to make his work known. He sent copies of his articles to Paris, which in the late eighteenth century was full of mathematicians: Lagrange, Carnot, Monge, Vandermonde, Laplace, Legendre, Lacroix and Fourier, not to mention Condorcet and Delambre.

Abel sent his articles to the Paris Institut, and to Cauchy, Legendre and other French mathematicians he was certain would recognize the value in his work. Then he waited.

Holmboe was hard at work in his office in the University of Christiania when the porter knocked and handed him a letter dated 'Froland, 6 April 1829'. There was only one line: 'Niels Henrik Abel died today at four o'clock in the afternoon.'

Abel had not lived long and his work as a mathematician was almost unknown. He would never know that the University of Berlin had just sent a letter offering him a post as professor. And at the *Institut* things were looking up.

Mr Ruche decided he would pay the Institut a visit. In the early afternoon, Albert picked up Mr Ruche and they headed towards the centre of Paris. As they crossed the Carrousel du Louvre, Mr Ruche glanced across at the pyramid. A group of Japanese tourists dressed in coats and woolly hats was crossing the square towards the pyramid, which looked like a crystal of ice. Around it, the water in the fountains was frozen.

Albert dropped Mr Ruche on the Quai du Louvre near the footbridge called the Passerelle des Arts. Mr Ruche rolled the wheelchair onto the bridge. The river below was a shimmering blue-grey. The leafless trees stood like sentinels along the riverbanks, and the pale sun seemed to warm the scene. On winter days like these, thought Mr Ruche, Paris was more beautiful than anywhere else in the world. A car backfired, bringing Mr Ruche back to his own century with a bang. He wheeled himself slowly across the bridge. He was in no hurry – he had a date with the immortals at the *Institut*.

At the gate he showed his ID card and was given a pass. He made his way into the second courtyard where two porters helped him up the steps and left him in the great hallway, from which a small lift led to the library. The library was a long narrow room with large ornate oak tables in the middle. He found the two books he had come to study, and settled himself at a table.

The first book was Niels Abel's *Mémoire sur une propriété générale d'une classetrès étendue de fonctions transcendentes* ('Memoir on a General Property of a Very Large Class of Transcendental Functions'). This paper had been filed in the library, unnoticed for three years before it was presented at the *Institut* – one week after Abel died. The second book, *Recherches sur les équations algébriques de degré premier* ('Research on First-degree Algebraic Equations'), had been sent to the *Institut* a month earlier by a boy who was still at school. According to his teachers, Évariste Galois was 'always busy doing things which don't need doing', though one said of him, 'I think he has some little intelligence but he hides it so well I have been unable to discover it.' Only one seemed to understand the boy, noting that, 'He seems to be raging against silence!'

Mr Ruche noted that Cauchy received Galois's article and understood its importance immediately. Unfortunately, the day he was to present his report, he was ill, and though he recovered quickly, the article was forgotten. Mr Ruche could easily imagine the boy's disappointment when he discovered that his work had not been presented to the *Institut* and the paper itself had been mislaid. But Galois simply went home and rewrote the article from memory.

In 1830, he submitted his article *Mémoire sur les conditions de résolubilité des équations par radicaux* ('On the Resolution of

Equations Using Radicals'), hoping to be included in the Grand Prix of mathematics. This time, another mathematician, Joseph Fourier, was to present the article but Fourier died in bed three days before and the article was not presented or submitted to the competition. A year later, Galois returned to the *Institut* a third time to depose his article. He received a response.

Poisson, who was responsible for a nice little theory of probability, read the article and replied, 'We have made every effort to understand the theory of Monsieur Galois. His reasoning, however, is not sufficiently clear nor sufficiently developed for us to judge its exactitude and we will not be able to do so in the report.'

Mr Ruche thought Poisson should at least have tried, and decided that he would check Grosrouvre's index cards to see what his friend had to say about Galois. It was 5.45 p.m. on the curious clock at the far end of the library. Built in the ninth year of the Republic, it had two faces. The upper clock face told solar time, the lower one civil time, and it showed both the Gregorian and Republican calendars. Mr Ruche tidied up his things and left. On his way out, he remembered that opposite the library door there should have been a statue of 'Voltaire, naked, at the age of 76'. It was missing. Mr Ruche asked a porter where it was, and was told that it had been replaced by a cenotaph to Mazarin.

'I suppose they thought an empty tomb to a dead general was better than the sight of the ageing flesh of a philosopher!' said Mr Ruche, smiling, as he left the library.

When he got back to the rue Ravignan, Mr Ruche recounted what he had learned to the children. They were indignant at the injustices both Abel and Galois had suffered and, that evening, the twins decided to do a little more research. They studied separately and met in Lea's room at midnight. Lea read what she had found:

My Dear Son,

This will be my last letter to you. By the time you read these lines, I shall be dead. I do not wish you to despair, nor that you be sad. Try to resume your normal life as soon as you can. I know that it will be difficult for you to forget a father who has also been a friend to you.

Her voice was almost inaudible. Jonathan sat beside her, staring out at the night sky, listening to his sister read the painful story Galois's father told his son. The letter concluded:

> *You will be a mathematician, my son. But even mathematics, the most noble and the most abstract of sciences, has its roots firmly in the ground. Even mathematics will not deliver you from your suffering and those of others.*
>
> *Fight, my child, fight courageously as I have failed to fight. I hope, before you die, that you may hear the bells ring out for Liberty.*

Lea's hands were trembling as she put down the letter. Galois's father had written it on the day he committed suicide.

Jonathan, who had done some research on Galois, shuddered when he thought how the son's life was mirrored in this letter. It was as though his father had mapped out his life for him before he died. He told Lea what he had found out.

'In 1830, fifteen years after the restoration of the monarchy in France, there was an insurrection in Paris. Galois wasn't involved because he was a pupil at a boarding school at the time. He made up for it later, though.'

Jonathan took out a piece of paper and unfolded it. It was a police report. Jonathan read it aloud:

'Galois took part in many of the riots in Paris. At a public meeting of the People's Friendly Society, he tried to incite insurrection, shouting "Death to the ministers!" He enlisted in the Royal Guard and spent the nights of 21 and 22 December 1830 trying to convince the artillery to turn over their guns to the people. On 9 May 1831, at the Republican banquet, he held a dagger in one hand and proposed a toast, "to Louis-Philippe".

'Character: in conversation, he is sometimes calm and ironic, sometimes passionate and violent. He is apparently a mathematical genius, though is not recognized by other mathematicians. He has no relations with women. He is one of the most dangerous of Republicans: brave, fanatical, extremist. He is naive and knows little of life, so it is easy for our men to gain his confidence.'

'Galois did have "relations with women" ', said Lea. 'I read about it. He fell in love with a woman, but she rejected him. The stupidest thing is that another man, who was also in love with her, challenged Galois to a duel. Galois knew he didn't stand a chance, so the night before the duel he wrote a long letter to Auguste Chevalier about his work.' She leafed through the book on her knees and began to read:

'...recently, I have been trying to discover the a priori relationship between quantities and functions, what changes can be made to them or what quantities can be substituted for the given quantities. In doing so, one quickly realizes that many of the expressions are impossible. But the terrain is vast. I have not had enough time and my ideas are not sufficiently developed. So often in my life I have advanced propositions of which I was not entirely sure, but everything I have written on this subject I have been thinking of for more than a year. It is important that I do not make a mistake lest people think I have stated theorems for which I do not have a complete proof.'

'As dawn approached', Lea said quietly, 'Galois finished the letter, enclosed the article he had been writing and left the room with his second and his witnesses for the duel.'

The following morning, Mr Ruche went to the Rainforest Library. He was still impressed every time by the sheer number of books in Grosrouvre's magnificent library. No one would ever suspect that such a treasure was hidden in this courtyard, he thought, which was just as well because Mr Ruche could not prove that the books in the Rainforest Library were not stolen. Grosrouvre had sent him no papers or invoices. There was the letter, of course, but that didn't prove anything.

He wheeled himself to Section 3, and located Galois – the young revolutionary who despised the aristocracy – sandwiched between a baron, Joseph Fourier and a prince, Carl Friedrich Gauss. In mathematical terms, he was in very good company.

Mr Ruche took down the book and settled himself at his desk. He decided to look back over his notes before going any further. He opened his notebook and read through the stages that mathematicians had passed through.

Mathematicians began by trying to find out whether an equation of a specific type had a possible solution. They did this by trying to solve it.

Then they discovered that some equations had multiple solutions. This raised another question: how many roots might an equation have? Was there an upper or a lower limit? They discovered the answer: an equation of degree *n* had exactly *n* solutions. This was the fundamental theorem of algebra.

At the same time, they tried to discover whether equations could be solved using radicals, and noted the formulas for equations of the first four degrees. It took another three centuries before Abel proved that a fifth-degree equation could not be solved using radicals. Then Abel and Galois proved that this was true not only of fifth-degree equations, but of all equations of higher degree.

It was Galois who had finally provided the proof.

Stating that no equation greater than the fifth degree can be solved using radicals, Galois wondered whether there was a way of finding out a priori whether a specific equation could be solved using radicals. Was there a specific criterion? There was – and Galois discovered it.

Perhaps it was his method that Poisson had failed to understand, thought Mr Ruche. He took out Grosrouvre's index card from Galois's *Complete Works*. It began with a phrase of Galois's:

'All geometry tends towards elegance.'

Mr Ruche stopped for a moment. He had always thought of elegance as one of the most beautiful aspects of knowledge. This sentence was written by a young man who would make geometry his life's work. Mr Ruche read on:

Rather than consider the solutions to individual equations, Galois considered them as a whole. Then he studied them to see what effect certain transformations, or *substitutions*, would have on them.

In his short burst of work, Galois solved the problem once and for all. But the way in which he did so opened up a whole new area in mathematics, and the methods he used would change for ever the way mathematics was done.

Algebra changed completely after Galois. He was interested not in numbers, or even in specific functions, but in structures. He was interested not in things individually, but in how they were linked together by their structure to form a whole.

These objects related by their structure are called a *set*, which would

> become perhaps the most important mathematical theme of the twentieth century, the focus of what became foolishly called 'new maths' – as though maths in every age were not new maths!
>
> N.B. Defining the structure of a set involves being able to identify the differences between two things which are not the same. It means breaking the homogeneity of elements within a set.

Mr Ruche liked this last note. It was in ways such as this that maths and philosophy came together, that he and Grosrouvre could meet on equal terms. Grosrouvre's card continued:

> The very newness of Galois's ideas lessens the criticisms we can level at his contemporaries. We can reproach them for not understanding his work, but we cannot say that they did not make a valiant attempt. Galois, however, paid the price for being so far ahead of his time. He simply couldn't wait for other mathematicians to catch up.

When Mr Ruche closed Galois's *Complete Works*, he remembered something that Cardano had written: 'Ensure that your book fulfils a need and that fulfilling it betters you. Only then will you have succeeded.' In this sense, the book Mr Ruche placed back on the shelf between Fourier and Gauss was a success. It had resolved one of the fundamental questions of algebra. He switched off the lights and left the studio.

He crossed the dark courtyard, trying to absorb all he had read. He wondered whether there was another way of solving algebraic equations – a method that mathematicians in Galois's time would have understood? In the field of mathematics, where proof is everything, Galois's tragedy was that he had provided proofs to confirm his assertions, but could find no one who understood the proofs, leaving him alone with his certainty. Mr Ruche shivered, and wheeled himself back to his room.

The following Sunday afternoon, everyone gathered in the living room to take stock. It was a dismal day – so dark that they had to turn on the lamp. Sidney dozed on his perch near the radiator. Lea had made tea for Mr Ruche and coffee for everyone else.

'All right', Perrette began, 'it all started with Tartaglia wanting to keep his work secret and then confiding in someone who was only pretending to be his friend.'

'If he really wanted to keep it secret, he shouldn't have told anyone', said Lea.

'He did intend to publish them', said Jonathan. 'He wasn't trying to keep them secret for ever.'

'Except that by the time he got round to it, it would have been too late', said Max, 'since he died before he could publish them.'

'Yeah, but how was he supposed to know that would happen?' asked Jonathan.

'Too bad', said Lea. 'It's his own fault that the guy who found out the secret got the credit.'

Perrette thought for a moment. 'And the story ends with Abel and Galois. What happened to them? They both tried their best to get their work published and understood. And it did them no good at all – at least in Galois's case. Maybe that's what Grosrouvre is trying to tell you, Mr Ruche. Maybe that's why he decided to keep his work secret – because if he had published his proofs, no one would have understood.'

Mr Ruche listened carefully. 'You're probably right', he said. 'After all, he was an old man living alone in the rainforest, sending off proofs to great mathematicians. They'd probably end up in the bin.'

'I think there's something else', said Jonathan. 'Tartaglia wanted his work to remain secret, but it was published anyway. Galois wanted to make his public, but it remained secret.'

'What do you think it means?' asked Perrette.

'That nothing turns out the way you expect', suggested Lea.

'The way you expect it to or the way you want?' asked Perrette.

'The way you want', replied Jonathan.

Sidney said nothing.

Max said, 'Grosrouvre seems to have come off best. He wanted to keep his work secret, and it stayed secret.'

'So far', said Lea.

Jonathan frowned. He didn't agree with Max at all. He took a piece of paper from his pocket: 'I wanted to read you something Galois wrote: "Egotism will no longer govern the sciences when groups of people study together. Instead of sending work to the Academy for deliberation, people will rush to print with the smallest observations, regardless of how little they add to the debate and add to their work..."

'And there was something else: "As a young man who has been twice rejected by them, I still dare to write. In this, I must be dedicated, for I expose myself to the cruellest of punishments: the mockery of fools. This is why I have now scaled every obstacle to ensure my work is published. So that, before I am dead and buried, those friends I have made will know that I am truly alive."

'He wrote that after his first two articles had been lost. Even then, he was against keeping things secret. What Galois is saying is that Grosrouvre is an egotist, and I'm afraid I have to agree', said Jonathan.

'If I were Galois...' Lea began, but she didn't get to finish her sentence as everyone dissolved into laughter.

'Yeah? What would you have done?' asked Jonathan, smiling.

'I would have asked my big brother to go round and beat them up!'

'And I'd have been more than happy to do it, too!'

'I'm not sure violence is the solution', said Perrette.

'I'd go mad trying to remember everything', said Lea, 'if every time I wrote something down it went missing.'

'What did you say?' asked Mr Ruche.

'Well, didn't you say that the articles he handed in to the *Institut* had been lost three times?'

'Trying to remember everything – what did Grosrouvre say about his friend? That he had an excellent memory, so that if his proofs were lost...'

Perrette jumped up, excited. 'Maybe I'm reading too much into this, but didn't Galois have a loyal friend too? What was his name?'

'Chevalier', said Lea. 'Auguste Chevalier.'

'And the night before the duel, he wrote him a letter explaining everything that had happened. He explained about the duel. And he confided his work to him.'

No one had noticed this obvious link. The night before his death, Grosrouvre, too, had written a letter, to Mr Ruche.

Mr Ruche nodded, a little confused. 'I'm not sure I was a loyal friend. An old friend, certainly. And he doesn't confide his results to me in the letter. That's the difference.'

'They're not the same at all', Jonathan protested angrily. 'One of them is a young guy about twenty, the other one is four times his age. One of them is a genius and the other one...'

'One of them was recognized as a genius forty years after his death', corrected Perrette.

'Well, we'll just have to wait forty years to find out about Grosrouvre.'

'I think you'll be waiting without me', said Mr Ruche softly.

When the twins had left, Mr Ruche asked Perrette, 'Do you know why they're so angry about this?'

'I think so', she said. After a minute, she continued: 'They've never liked secrets. One thing surprised me in the story they told. I assumed that the duel was with a royalist – but it wasn't. It was a duel against a friend – a Republican officer.'

'I don't see what you're getting at.'

'I'm not sure. You know your enemies are out to get you, but what if Grosrouvre's enemy was really a friend?'

Perrette had mentioned that friends might have killed Grosrouvre when they had all been discussing al-Khayyām's story of the 'three friends'. Galois was up against a friend and an officer, trained in using firearms. He didn't stand a chance, any more than Grosrouvre did against the gang.

'Maybe we're being paranoid', said Mr Ruche after a while.

'I don't think so', said Perrette quietly.

For a fraction of a second before she fell asleep, Lea retraced the paths of the blade that had disfigured Tartaglia and the bullet that had killed Galois. Next door, lying on his bed beneath the skylight, Jonathan was reliving the duel. The two white handkerchiefs on the grass, twenty paces apart. The pistols chosen, Galois and his old friend walked away from each other, turned, and fired. Galois would hear his second say 'You have one minute to get up', and then no more.

Lying on the grass, railing against the silence.

CHAPTER 18

Between the Infinite and the Void

Descartes, Pascal and Fermat; analytical geometry and number theory; Newton and Leibniz & differential and integral calculus

As he passed the florist's, Mr Ruche stopped to smell the mimosas. They always reminded him of the hills around his old home in Bormes. In the spring, they were ablaze with mimosas in flower, and the heady scent was the first sign of nature bursting out once again. Mr Ruche would gladly have gone south to see them in bloom, but instead he bought a bunch of flowers and gave them to Perrette, who put them by the till in the bookshop. For several days they brightened up the shop.

Mr Ruche was exhausted after his work on algebraic equations and decided that he needed a break. He would take a few days off: no library, no Manaus, no Grosrouvre. He phoned Albert, and they arranged to drive into the country and spend the day together. It wasn't warm enough for a picnic, but they would find a nice inn on the way.

They headed west out of Paris and turned off the motorway at Mantes-la-Jolie, heading down towards Rolleboise, where they took the road which ran along the river. The road left the bank and climbed to a high plateau overlooking the Seine. They drove down through a small forest. The road swung out of the woods suddenly, and they came to a spectacular view by the river's edge. There, on the bank, was a small bar and restaurant with a thatched roof called The Riverman's Rest.

They went inside and looked around. There was not a soul in sight. By the bar, Mr Ruche spotted a sign which he read aloud: '*Curva sequana, mens recta* – The river is curved, but our will is straight.' As he did so, a young man appeared behind the bar and offered them a menu.

They sat by the window overlooking the river. On the far bank stood a beautiful church. Mr Ruche thought he recognized it, though he knew he had not been here before. Albert said that he sometimes recognized places he had never been to. *'Déjà vu'*, said Mr Ruche. They ate snails in garlic butter and farm-fresh chicken washed down with a rosé wine. It was warm and bright and behind the tall windows they felt like flowers in a greenhouse.

The next day, Mr Ruche went to the bookshop and looked in all the books on Impressionist painters. Finally, he found what he was looking for: the church he had seen by the river was the Église Vétheuil. Monet had painted it many times from the boat-builder's yard exactly where the restaurant now stood.

Happy with his discovery, he returned to his desk in the Rainforest Library to work on the next mathematician. He wondered idly what he used to do before Grosrouvre's letters arrived, setting him off on his mathematical studies. How had he managed to pass the days without dying of boredom?

The next name on the list was Pierre Fermat. He was significant in Grosrouvre's work as it was one of his theorems that Grosrouvre claimed to have proved. Without thinking, Mr Ruche wrote πR, the way Grosrouvre had written his name in the first letter. Then, underneath, he wrote 'Fermat' and drew a circle around it:

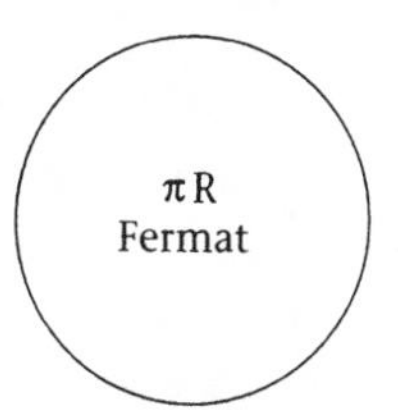

He wheeled himself to Section 3 of the Rainforest Library, Mathematics in Western Civilization, 1400–1900. He found Fermat's *Œuvres complètes* (Complete Works), in five volumes, and opened volume one to look for Grosrouvre's index card. There was a whole sheaf of cards.

Mr Ruche settled himself at his desk. As he began to take notes he quickly realized that any similarities between him and his namesake Pierre Fermat ended there. Fermat had five children and Mr Ruche's bookshop was a poor substitute for Fermat's work at the *Parlement* of Toulouse, where he was Advisor to the Publishing Council and to the Research Department. Fermat did more than propose his famous theorem, Mr Ruche discovered. In fact, it is probably the least important aspect of his work. As Grosrouvre had noted,

> Fermat developed the modern theory of numbers, laid the foundations for probability (with Pascal), created analytical geometry independently of Descartes (who was also working on it), and pre-empted the work of Leibniz and Newton on the calculus of integration and differentiation by some years. Like Viète, whose notation he adopted, Fermat was an amateur; in fact, he came to be known as the 'Prince of Amateurs'. He published nothing – most of his work was communicated in letters and remained in manuscript form until after his death.

Mr Ruche leafed through the *Œuvres*. The greater part of the five volumes consisted of letters. Letters to the greatest mathematicians and intellectuals of Europe: Mersenne, Carcavi, Frénicle, de Bessy, Pascal, Descartes...It was clear to Mr Ruche that he could not tackle Fermat without studying Pascal and Descartes. He knew their philosophical works very well, and was happy to have an opportunity to look into their mathematics. He went back to the index cards.

> Fermat's work was about continuity. There are no great declarations in his letters. He was not, like Descartes, intent on revolutionizing mathematics, though he did change it radically. Building on the work of Apollonius, he founded analytical geometry; continuing the work of Diophantus, he developed number theory; and following on from the work of Archimedes, he laid the groundwork for the integral calculus. Fermat was a weathervane for what was happening in mathematics in the middle of the seventeenth century. His work went off in four separate directions, each opening up a vast field of work.

Mr Ruche began to wonder how to approach 'πR Fermat'. He picked up the paper on which he had written Fermat's name in a circle and began to write down what he had just learned.

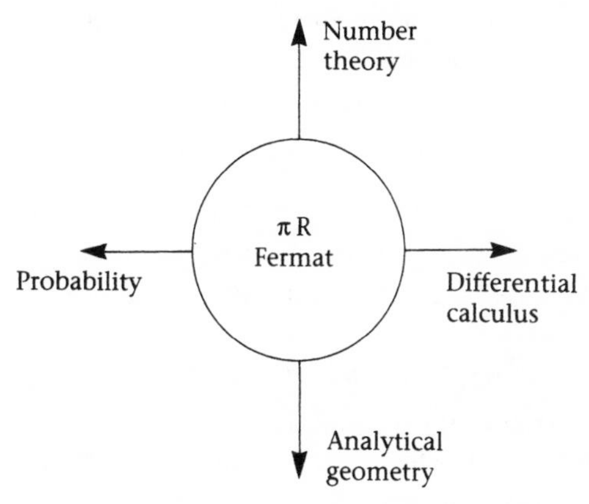

Mr Ruche knew that the only way to deal with Fermat was to take each of these directions on the weathervane in turn. He decided that he couldn't tackle this alone, so he asked the twins to help. When he showed them the weathervane of Fermat's work and asked which directions they would like to follow, the twins opted for the same direction. Mr Ruche had hoped they would take a different one each. Without a moment's hesitation, Jon-and-Lea picked west: Probability. The Rainforest Library door closed behind them, and Mr Ruche was left with three paths still to follow.

The door opened again and Lea came in. Mr Ruche was impressed: it seemed as if the kids were being brave enough to take on a second subject. But Lea walked straight past him to the shelves of the Rainforest Library, took down some books on probability by Pascal and went out, closing the door behind her.

Mr Ruche thought of the mimosas and chose south:

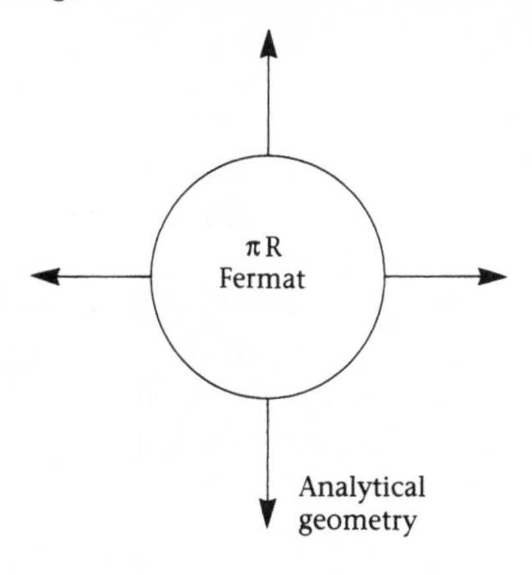

> The principles of analytical geometry can be summed up in one sentence: from the equation of a curve all the information relating to that curve can be found. Fermat and Descartes discovered this independently of each other and it was called *coordinate geometry*.

Mr Ruche immediately understood what it was about and was surprised that he had never heard of Fermat at school. Descartes, on the other hand, he had heard all about. He had his own adjective: *Cartesian*.

He began drawing. Years of maths at school had taught him to draw a horizontal line for the x direction (the x axis, or abscissa) and then a vertical line for the y direction (the y axis, or ordinate). At the intersection, he wrote a large O (the *origin* of the coordinates):

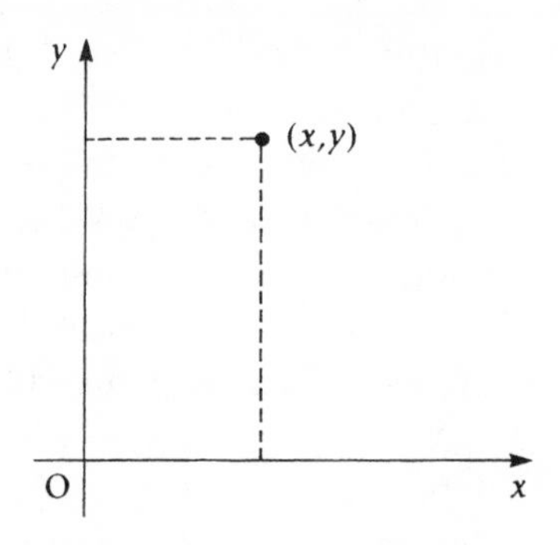

He suddenly remembered playing as a boy:

'(A, 8).'

'Hit!'

Mr Ruche had played battleships all the time – it was his favourite game. All you needed were two pencils, some paper and a rubber. The paper had to be squared, not lined, so Mr Ruche used to tear it out of his maths exercise book. It was strange that a boy who couldn't swim, who got seasick just standing on a pier, could spend days reliving the battle of Trafalgar. Whether he was playing a filibuster in the Dutch fleet or an admiral of the Royal Navy, he fought across the seven seas – his cannonballs following Tartaglia's parabolas and crashing onto enemy ships. Hit!

As he daydreamed about his boyhood, Mr Ruche fell asleep. When he woke, he felt something stroking his head. Sidney was delicately grooming his hair. 'He's a strange bird', thought Mr Ruche. 'Charming, but strange.' Perched on the desk only inches from his face, Sidney watched him closely, his large, yellow-ringed irises fixed on the old man. Mr Ruche scratched the back of the bird's neck as he had seen Max do. 'You have to stroke him in the direction of the feathers', Max had insisted.

Then, picking up his pen, Mr Ruche wrote:

> Like ships at sea, the points on the graph are identified by their coordinates. How can you communicate the position to someone who cannot see it? The point on the graph identifies it like a name. Like people, points on a plane need something to identify them!

He returned to Grosrouvre's notes:

> The ordinate axis can be placed anywhere, nor does it matter what unit of measurement is used. Negative coordinates, however, were disliked at first, especially by Descartes, until the arrival of an Englishman, John Wallis. Like Viète, Wallis was a great decoder of secret letters.
>
> He had studied at Cambridge and lectured at Oxford, and sided with Cromwell and the Parliamentarians against Charles I. Wallis decoded many of the Royalist messages which fell into Parliamentarian hands. However, he was opposed to the execution of the king.
>
> He was a mathematician, logician, grammarian and doctor. He translated the works of Nasīr al-Tūsī and was interested in Euclid's fifth postulate. As a doctor, Wallis was the first scientist to publicly support William Harvey's theory that blood circulated in the body. He also opened the first school for deaf-mutes in Britain.

Max had never gone to a special school, thought Mr Ruche. Because he had not been born deaf, he was able to speak. His speech was slow, intense, each word, each silence carefully measured. And he had his own way of listening.

Mr Ruche went back to his work. Looking at his notes afresh, he suddenly saw the discovery by Fermat and Descartes more clearly: these simple axes had completely changed the nature of space. He wrote quickly:

Using the axes, geometry became algebraic – the point M became a pair of numbers (x, y). It was a revolution. The same was true for the geometric curve. The equation that defined the curve became its algebraic name and could be used to find any point on the curve. Knowing the equation gave you all the geometrical information about the curve.

Mr Ruche sketched a graph with a curve and a point. This was how an equation was represented, and any point (x, y) which was a single instance of the equation:

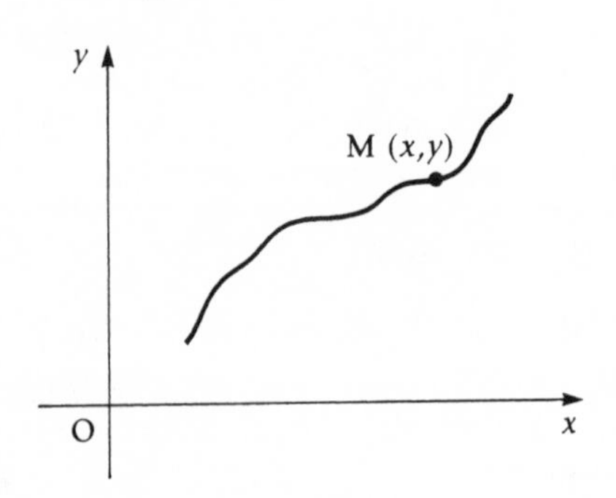

Mr Ruche congratulated himself, and went back to Grosrouvre's cards:

Fermat had developed his system to open out geometry to the riches of algebra. For him, geometry remained at the heart of mathematics. For Descartes, on the other hand, algebra was a much greater science than geometry, which he thought of as something solely for calculation.

The Greeks had developed the science of geometry and now, in the seventeenth century, they entered the realm of algebra. Descartes placed algebra on the throne from which geometry had once reigned supreme.

Descartes had written a lot but, Mr Ruche discovered, there were few of his books on the shelves of the Rainforest Library. There was a hardback copy of *La Géométrie*, and copies of *Discours de la Méthode* and *Regulae ad directionem ingenii* ('Rules for the Guidance of the Mind'). Mr Ruche knew the opening lines of this last book by heart:

> Actors upon a stage wear masks to hide their blushes. As I set foot upon the world stage for the first time, having hitherto been a mere spectator, like them I advance masked.

He took down the copy of *La Géométrie* from the shelf. It was a remarkably slim volume, but in these few pages Descartes proposed a five-point programme which anyone faced with a geometric problem should follow. Mr Ruche wrote it down:

1. Consider the problem to be solved. This allows one to *analyse* it (to move from the unknown to the known).
2. Deconstruct the problem into its component values and classify them as to whether they are known or unknown, then attribute a letter to each part.
3. Establish a hierarchy of importance, without distinguishing between what is known and unknown.
4. Attempt to express one of the values in two separate ways. One expression equals another, and an equation is created.
5. Try to find as many equations as there are unknown values. If you cannot do so, then the problem is not completely soluble.

Mr Ruche was impressed. He could see that, using Descartes's method, analytic geometry was efficient. In his *Discours de la Méthode*, Descartes stated, 'A method is essential if one is to investigate the truth of things.' For Descartes, Mr Ruche realized, algebra was not a science; it was simply a 'method', from the Greek *meta-hodos* – *hodos* meaning 'the way' – a method which, if followed, was a route to an end.

In their eyrie under the skylights, Jon-and-Lea were heading 'west' on Fermat's weathervane, although they also spent long hours dreaming of true west – travelling across the Atlantic and up the Amazon to Manaus. Manaus now was not the city it had been. It was no longer the glorious centre of the world's rubber production, as Jonathan found from the book he was reading:

> Henry Alexander Wickham was preparing to leave Manaus for the last time. He made his way through the streets to the harbour, where his

ship was docked. He marvelled as he passed the chateaux, which had been brought stone by stone from Europe, the covered market, built by Gustav Eiffel, the streets paved with stones from Lisbon. Electric trams, the first in South America, passed him as they carried people to and from their work. They had electric light here, telephones, and an opera house where Caruso himself had come to sing. Wickham was about to destroy the city with a single act.

His heart beating fit to burst, Wickham calmly answered the customs officer inspecting his ship. 'I am taking specimens of rare plants to England. I shall plant them in the Botanical Gardens at Kew.' Reassured, the customs officer disembarked.

Wickham rushed to the hold and stared at the carefully packed baskets. In them was a treasure that would be England's fortune and the ruin of Manaus. These were not rare plants, but seeds – 70,000 of them, carefully packed in dried banana leaves – and Wickham had no intention of giving them to Kew. They were seeds for *Hevea braziliensis*, the best rubber plants in South America. It was illegal to export them, but Wickham's bluff had paid off.

Decades later, transplanted into the Malayan forests, Wickham's seeds blossomed into vast plantations with a huge output of rubber. Manaus was ruined. The people deserted it, and the city crumbled.

Lea shook him by the shoulder. 'I'll read you my notes', she said. While Jonathan had been dreaming of Manaus, she had been reading up on the work of Pascal, who was credited with developing probability.

'OK, biography first: Blaise Pascal's mother died when he was three, leaving his father and two sisters. Jacqueline, his older sister, became a nun, and Gilberte became Mrs Périer. Blaise Pascal was a child prodigy and his father, Étienne Pascal, like Mozart's father, insisted on teaching the child himself. Blaise never went to school and didn't have any friends to play with.'

'Can't have been good for him psychologically', remarked Jonathan.

'It doesn't seem to have done him any harm. Étienne was a mathematician himself. He discovered a curve – called the limaçon of Pascal – which was a conchoid, and a specific instance of Descartes's ovals. So, you see, it was also a trisectrix! Hey! Are you listening, or what?'

'I'm listening, but I'm not really following.'

'Pascal's father stopped him from studying geometry because he was worried it might cause mental fatigue. So what happened? Blaise studied geometry in secret. He was terrified that his dad would find out, but that just made it more exciting. By the time he was twelve, he'd "discovered" Euclid's Proposition 32 all by himself – the sum of the angles in a triangle equals 180°. He didn't even know who Euclid was. (At least, that's what his sister says.) When his father found out, he was so happy that he gave Blaise all thirteen volumes of Euclid's *Elements* as a present.'

'Some present. Abel was studying maths at twenty-one, Galois at eighteen and Pascal at twelve – it looks like a number sequence heading for 0 to me!' quipped Jonathan, a little irritated by all these young geniuses. He was angry that he was now seventeen without a single brilliant idea to show for it.

'Well, Grosrouvre didn't start working on his proofs until he was sixty', countered Lea, 'so if he did solve them he's doing well. Maths is difficult enough – but it must be worse if you're a pensioner. I read somewhere that a mathematician's best work is over by the time he's twenty. There's not much chance he will accomplish anything later.'

'Not much chance?' said Jonathan. 'But how much chance? That's what probability is. Gymnasts are the same – past it by the time they're twenty.'

'Same difference, really – maths is just mental gymnastics, and little Blaise was a pretty good gymnast', said Lea. 'He'd written *Essai pour les coniques* ('Essay on Conic Sections') by the time he was sixteen. There's a theorem in it that created a lot of interest when it was published. You take a six-sided polygon...'

'A hexagon', interrupted Jonathan. 'Use the proper name!'

'Smart-arse! OK, take a hexagon inscribed in a circle. There are six sides, so there have to be three pairs of opposing sides. Pascal discovered that where they intersect, the three points are aligned.'

'So?'

'And he proved that it works for any conic section – ellipse, parabola, hyperbola...'

'Do you really understand everything you're saying', asked Jonathan.

'About half of it.'

'Then why are you telling me?'

'Because I don't want you to die stupid.'

'Listen, Pascal called the hexagon the 'mysterious hexagram', and he called his theorem the cat's cradle.'

'Yeah, well I'm tired...' said Jonathan, curling up under the duvet, '...and you know what killed the cat.'

Two seconds later, Jonathan was purring gently.

Lea remembered that for every new mathematician Mr Ruche found somewhere new to go: the National Library, the IMA, the *Institut*. She decided she should find somewhere to study Pascal.

The next day, she took her brothers on a mystery tour. Max brought Sidney, who hadn't been out for some time. They walked along the Grands Boulevards towards the city gate at Saint-Martin. As they walked, Lea talked about Pascal's theory of wheelbarrows. Jonathan and Max tried not to giggle when Lea solemnly informed them that Pascal's improvements are still used in modern wheelbarrows.

They arrived at the National Conservatory for Arts and Crafts and wandered round the old refectory, which had been converted into a library. Afterwards, they went to the church where they were disappointed not to find a wheelbarrow, but there was another of Pascal's inventions, next to Foucault's pendulum. Jonathan knew about it, having read Umberto Eco's novel.

Sidney, thrilled by the vaulted ceilings, flew from Max's shoulder and started looping the loop, to the delight of the other visitors, until the museum attendant came over and insisted that Sidney stayed perched on Max's shoulder.

Lea decided it was time to get down to work and took on the role of guide.

'Pascal's father, Étienne', she told them, 'was a tax-collector in Normandy. It was a well-paid job – the more you collected, the more you earned, so you were motivated to do well. Blaise loved his father dearly and invented for him a calculating machine – the *Pascaline* – which they called an "arithmetic machine".'

The machine was in the case in front of them. It was a wooden box with six cogs, each with ten golden spokes, one for each of the ten numerals.

'It's a simple machine, but one you could count on', commented Lea.

'Good pun. "One you can count on"!' Jonathan said sarcastically. 'It all adds up!' Then, seeing the surprise on Lea's face, 'and it wasn't even deliberate!'

'The important thing about calculating using a machine is what you do when you come to 9 and you need to carry 1', said Lea. 'Pascal devised a little mechanism which automatically carried over. He set up a small workshop, drew up the plans for his machine and patented his design. He manufactured about fifty *Pascalines* and sold them for one hundred *livres* apiece. That was a lot of money back then. In his book *Pensées*, Pascal said that the machine was closer to human thought than anything in the animal kingdom. But he also said that nothing it could do indicated that it had a will, as animals do.'

'What do you think?' Max asked Sidney. Clearly the parrot couldn't care less what Pascal thought animals could think.

On the other side of the city, outside Roissy airport, Albert was about to light a cigarette when a stocky man leaned through the driver's window and asked, 'Could you drive me to Paris?'

'Where did you fly in from?'

The man was puzzled, but replied, 'Tokyo.'

'Sorry, mate, I'm not interested.' To the man's astonishment, the taxi pulled away, stopped a little further down the road and picked up another passenger. Stunned, he walked back to the rank and waited. When his turn came, he climbed in and the driver took the motorway north towards Paris. It was raining softly.

Leafing through his briefcase, the man had a nagging suspicion about the taxi driver who had refused him. He shuffled through his papers, opened a folder and took out a photograph.

'Good God!' he shouted.

'Something wrong, mate?' asked the driver, looking at him in the rear-view mirror. The man stared intently at the photo. There was no doubt about it. The man in the photo, standing next to the boy with the parrot, was the taxi driver who'd refused his fare. It was amazing – he was even wearing the same cap.

He leaned forward and said to the driver, 'I have to catch up with the cab that drove off just before we did.'

'Sorry, mate, it's gonna be difficult to go any faster.'

'It's very important. If you catch up with him, I'll make it worth your while.'

'What kind of car did you say it was?'

'A Renault 404.'

'You didn't notice what cab company he was with, did you?'

'No, sorry.'

'It's not gonna be easy, I can tell you that. Just look at all the cabs.'

They were surrounded by taxis, but not a Renault among them.

'You sure it was a cab, guv?'

'What kind of idiot do you take me for? Of course it was a cab.'

'I mean, an official taxi. Was there a sign on the roof?'

'Yes.'

'Was there a little sign in the back like that one?' the driver pointed to a small pink disc in the windscreen. 'You can't really see it from inside the cab, but that's the only way to know for sure he's on the official register. I'm just saying, 'cos you get a lot of guys driving fake cabs nowadays. They go out and buy taximeters on the black market.'

'You said there was a register of taxis. Where is it kept?'

'At police headquarters.'

The Tall Stocky Guy knew they would never catch up with the Renault now, but he had a lead. He would find the register and track down the taxi. The Boss would be happy. They had two leads now – the photo and the taxi.

Back in the Rainforest Library, Mr Ruche went back to his weathervane of Fermat's disciplines. He looked over the diagram and took stock. Jon-and-Lea had taken the west, and he had taken a short walk south through the world of analytical geometry. Only the north and east remained. He was certain that north was the direction Grosrouvre wanted him to take, so he decided to leave it until last. First he would head east.

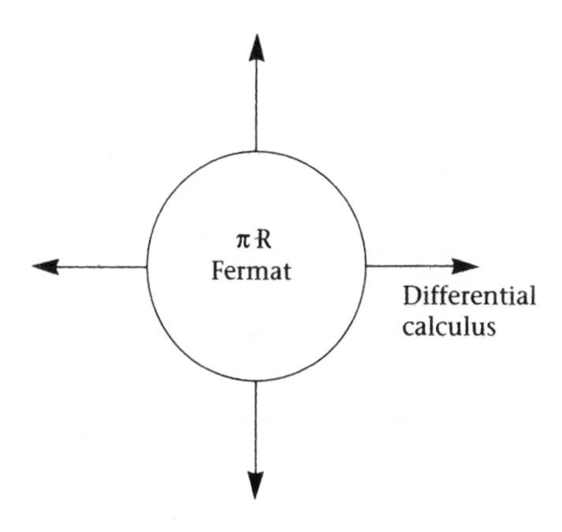

Mr Ruche put together a list of those who had contributed to the birth of differential calculus. It read like a *Who's Who* of seventeenth-century mathematics:

Italy: Bonaventura Cavalieri and Evangelista Toriccelli.
France: Fermat, Roberval, Pascal and Descartes.
Holland: Christiaan Huygens.
Germany: Gottfried Wilhelm Leibniz.
Switzerland: Jacques and Jean Bernoulli (Jacques invented the word 'integral').
Great Britain: Isaac Newton, Isaac Barrow, Christopher Wren, John Wallis, James Gregory, Brook Taylor and Colin Maclaurin.
The masters of this new discipline were Newton and Leibniz.

Mr Ruche drew a curve:

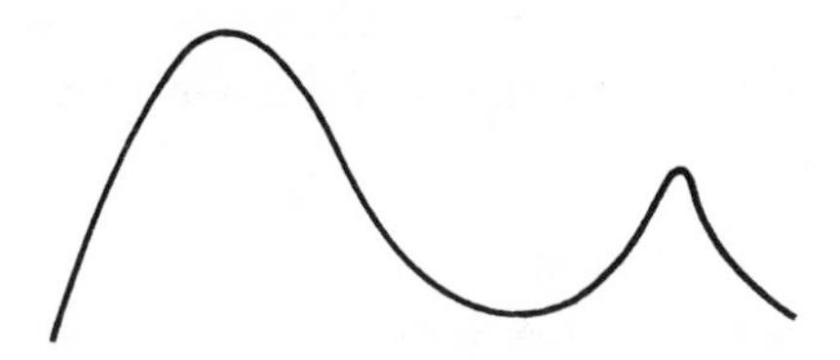

The first thing to do, he knew, was to mark the *maxima* and *minima* – the highest and lowest points of the curve. He also marked the *points of inflection,* where the curve changes from concave to convex.

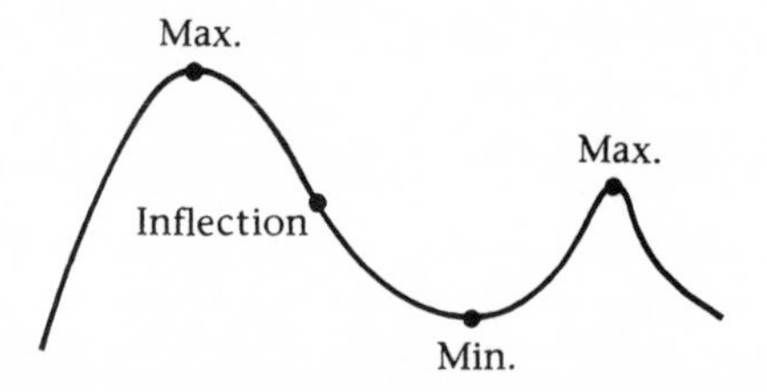

'What exactly are the "maximum" and "minimum" ', Mr Ruche wondered to himself. 'The terms signify the "extremities", but if you look carefully, the point just before is infinitesimally close, as is the point after.'

Fermat, he discovered, made this the basis for his research into maxima and minima, translating them from geometry into the language of algebra – into equations. Mr Ruche jotted down some notes on what he had been reading:

> 'Before' is an easy concept to define in maths, but 'just before'...how can one represent 'just'? The difference between a point and what comes 'just before' is tiny – a difference we can barely imagine. It is infinitesimal.
>
> *Differentiation* took hold of the imagination of the seventeenth century. In a society fascinated by microscopes, scientists were curious to take 'a closer look'. Until then, specific ideas sometimes led to universal ideas. Now, ideas of the infinitesimal might lead to universal ideas.
>
> What were these new infinitesimal things? Were they geometric values or numerical values like Fermat's? Leibniz thought they were useless fictions. But, like imaginary numbers before them, though no one knew quite what they were, they were made to work and they produced extraordinary results.

This was something that had passed him by as a student. While he had been studying for his degree in philosophy, his lecturers had touched on the subject, but he had never pursued it. Now, sixty years later, he was beginning to understand. He continued to write:

> Fermat's discovery: the smallest arc of a curve can be assimilated into the corresponding segment of the tangent.

Roberval's discovery: the direction of the movement of a point describing a curve is the tangent to the curve at that point. In fact, a curve was defined by the direction of the tangent – the form of a curve can be mapped out by a set of straight lines. Everything comes back to the relationship between the curve and the line.

At the time, the infinitesimal was known as the 'disappearing' and tangents were referred to as 'touchings'. These were the two key ideas. Newton described them as 'quantities which diminish not to the point where they have disappeared, nor to the point before, but to the point where they are disappearing'. It sounded like a poem.

And a tangent? If a line intersects a curve at points M and M', it becomes a tangent when those points are 'infinitely close one to the other'. A tangent did not intersect a line; it barely grazed it.

A tangent began as a line intersecting the curve, at points M and M', but as it is turned about the point M, it ends as a tangent, which does not intersect the curve but just touches it.

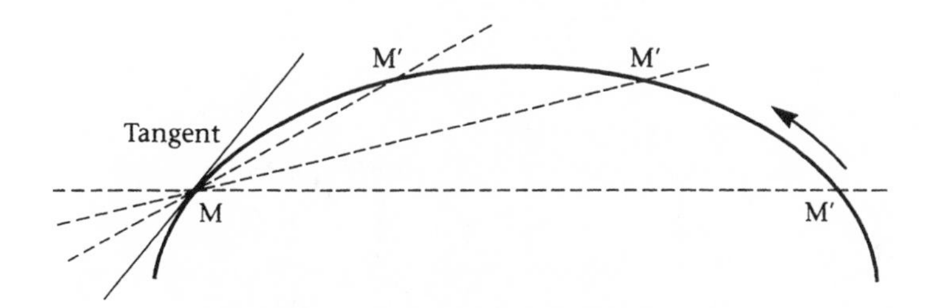

That much was easy. As for differential calculus, Mr Ruche had terrible memories of it from his schooldays. He thought he remembered that derivatives measured the specific variation of a function. He looked at Grosrouvre's notes:

Mathematically it is easy to find the variation of a function for a given range. *Derivatives* make it possible to find a precise value corresponding to a precise value for the variable, calculating the relationship between infinitesimal changes in the function and the corresponding changes in the variable as it tends towards 0. Take, for example, the formula which defines the derivative $f'(x)$ of the function $f(x)$. If Δx denotes the variation of x, and Δf denotes the corresponding variation of f, then

$$f'(x) = \frac{\Delta f}{\Delta x} \quad \text{as } \Delta x \text{ tends to } 0,$$

which we can write as

$$f'(x) = \lim_{\Delta x \to 0} \frac{\Delta f}{\Delta x}$$

Mr Ruche scratched his head. He still did not understand the formula. He could grasp the idea of 'tending towards a limit' – getting as close to it as possible without actually getting there.

From the idea of differentiation came the idea of *integration*. When measuring a curved surface, seventeenth-century mathematicians began to think of the area not as a whole, but as a series of bands placed side by side which together made up the whole. This area, then, was the 'sum' of an almost infinite number of narrow 'bands', each with a surface area that was almost non-existent.

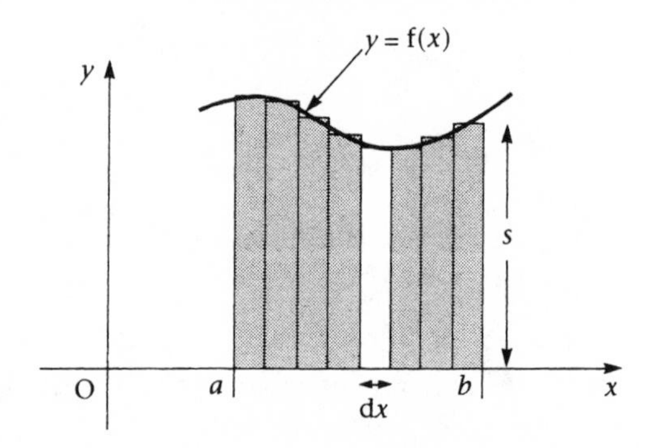

Mr Ruche reread what he had written: 'the "sum" of an almost infinite number of narrow "bands", each with a surface area that was almost non-existent'. It was difficult to imagine what the sum of an almost infinite number of elements might be – especially as the elements were infinitesimally small. Mr Ruche found it a strange concept: an addition that calculates not a finite number of finite numbers, but an 'infinity' of minuscule values, but ends up with a finite value. That was what 'integration' meant.

He decided to go over it again. He thought for a moment, and realized that integration was the sum of an infinity of 'infinitesimals', which produced a finite result. He felt he was beginning to get somewhere:

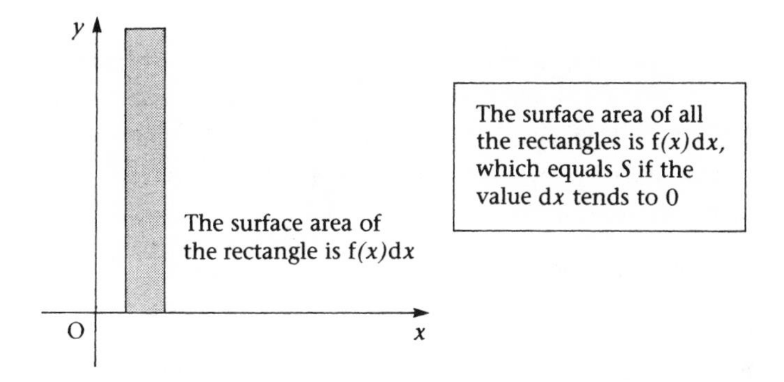

To symbolize this sum, Leibniz introduced an elongated ∫, which was the sign for integration – the sum of an infinite number of infinitely small rectangles:

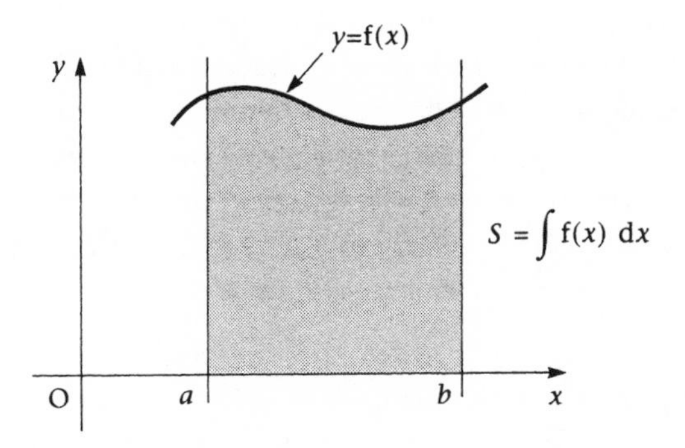

If a line is straight and its length finite, it can be measured easily, as can the surface it contains. A space defined by a curve is more difficult to measure. Integration used infinitesimally narrow bands to measure the curve, thus straightening curves, squaring surfaces and cubing solids, and making it possible to calculate lengths, areas and volumes.

To algebra, geometry, trigonometry, the other classical disciplines, a new discipline was added: *analysis*, which encompassed differential calculus and integral calculus. Mr Ruche went back to reading Grosrouvre's cards:

> Newton and Leibniz were the founding fathers of analysis. They made two important discoveries:
>
> 1) Determining tangents and calculating areas were simply different aspects of the same thing, and one could move from one to the other from the tangent to the curve, or from the derivative of the function it was derived from. Here was a single tool which was capable of performing very different actions:
>
> Measuring the length of a curve.
> Determining the area of a figure or the volume of a solid.
> Locating the maxima and minima of a curve, its tangents.
> Expressing speed and acceleration.
>
> 2) Newton and Leibniz developed rules for this new calculus. Finding the derivative, called differentiation, became an operation, not on numbers but on a range of variables defined by curves. It was an operation that could be done using a systematic algorithm.
>
> After centuries in which mathematics had four simple arithmetic functions and the extraction of roots, suddenly differentiation and integration were added. Together they were to provide a universal tool which would prove essential to the world of physics. Variations of all kinds could now be studied, and movement and acceleration, which had long been excluded from mathematics, could now be included. The static shapes of Greek mathematics began to move.

Mr Ruche noticed that, unusually, Grosrouvre had included two quotations on his index cards:

> I do not know what I may appear to the world; but to myself I seem to have been only like a boy, playing on the sea-shore, and diverting myself in now and then finding a smoother pebble or a prettier shell than ordinary, whilst the great ocean of truth lay all undiscovered before me.
>
> Isaac Newton

> Those who clearly see the truth of the geometry of indivisibles can admire the grandeur and power of nature, which, in this double

> infinity, surrounds us on all sides. They can learn more of themselves and their place between the infinite and the void of space, the infinite and the void of numbers, and the infinite and the void of movement and of time. In this, we can learn our true value and form ideas of greater worth than all the rest of geometry.
>
> Blaise Pascal

For a time, Mr Ruche sat listening to the wind in the courtyard, imagining himself on a deserted beach – a place beneath the infinite and the void. He dreamed of stretching out his arms to touch one and the other, to know his true value. He fell asleep in his chair in the Rainforest Library. All night he ran barefoot on the shore.

CHAPTER 19

A Narrow Margin

Pascal and probability; number theory & Fermat's Last Theorem

As she came out of the métro station at Barbès, someone handed Lea a small card:

> Monsieur Simakha, Medium and Clairvoyant
> One who has inherited great gifts of seeing
> There is no problem without a solution

Lea slipped the card into the back pocket of her jeans and headed towards the café on the rue Lepic where she had arranged to meet Jonathan.

'Chance would be a fine thing', said Jonathan cryptically, as he sat down opposite Lea on the terrace of the café.

'What are you on about?'

'Chance. Probability. You haven't forgotten we were supposed to study it? The trip to the National Conservatory was all very nice, but it didn't get us very far. Anyway – this is what I know. Probability is always somewhere between 0 and 1. A probability of 1 means something is certain. A probability of 0 means it's impossible. Between the two are all the possible shades of probability. From what I've read, probability is about making chance or luck mathematical. Pascal called it "the geometry of chance".'

'Making probability precise sounds like cutting off a bird's wings', said Lea.

'What are you talking about?'

'Well, on the way here I was wondering what the probability was of Max running into Sidney at the flea market.'

'Well, it can't be a probability of 0. Did you ever wonder what the probability was that we would be born twins?'

'Yes.'

'Well, I can't tell you exactly – but I'm sure we could find out. Anyway, this is what I've worked out so far. Probability began in the seventeenth century with a conversation between Pascal and his neighbour, Méré, who was a bit of a gambler. They were travelling together. At a staging post where the coach stopped for the horses to be changed, Méré talked Pascal into playing dice with him. When the coach was ready, they had to stop the game. Méré asked Pascal what they should do with the stakes. To divide them fairly, they would need to know the probability of each of them winning, but how would they go about finding it?

'Soon after, Pascal wrote to Fermat. Actually, I'm not sure it happened exactly like that', admitted Jonathan. 'But anyway, Pascal and Fermat wrote to each other about the problem, and the basis of probability was developed in their letters. Pascal also worked on combinatorics – the number of ways of calculating possible cases without having to count them one by one – arrangements, combinations and permutations.

'Pascal had a definition: "the probability of an event is the number of favourable cases divided by the number of possible cases." '

'So being a twin is a favourable case?'

'Well, I think it is...anyway, listen to the rest.'

The waiter came over to them. Jonathan ordered a glass of milk and Lea a black coffee.

'After they had studied games', Jonathan went on, looking at his notes, 'card games, dice, roulette – they started taking probability seriously. Do you know, they started working out tables for the probability of dying? They worked out the probability of a random person surviving. They were called mortality tables.'

'So we go from multiplication tables to putrefaction tables!' quipped Lea.

'You have such a delicate way of putting it.'

The waiter returned with their drinks. Lea pointed to the coffee:

'Black: impossible, white: certain. And between the two, all the possible shades of milky coffee.'

Jonathan went back to his notes to pick up the thread of what he was saying.

'You remember Mr Ruche telling us about the Bernoullis – there were loads of them. In two hundred years there were about ten, and they were all mathematicians. They weren't exactly a close family. Jacques, the eldest, hated Jean – if they both attended a lecture at the Academy, they usually ended up arguing. Anyway, Jacques wrote the first book about probability: *Ars conjectandi* – "The Art of Guessing". He died while he was finishing it.'

'I bet the mortality tables didn't predict that one!'

'Probably not – anyway, one of the other Bernoullis found the manuscript a couple of years after he died and published it. It had a huge impact on maths at the time. Bernoulli thought that things themselves are not uncertain. The uncertainty is in our heads: uncertainty is simply a lack of information. Or, as he put it, "Tomorrow's weather cannot be other than it will be in reality." '

'Well, if it's just information we need, how about this?' Lea took the card from her back pocket and, in a theatrical voice, read aloud:

'Medium and Clairvoyant. There is no problem without a solution.'

'That's exactly what Bernoulli says', said Jonathan. 'His aim, and I'm quoting here, is to "discover the general laws which govern what, in their ignorance of cause and effect, men give the names of chance and fate".'

'What about sudden impulses? What about my secret desires? What about...' – Lea was furious now – '...what about free will?' she shouted, knocking over her coffee and soaking her jeans. She yelped and shot up from her seat, almost knocking over the next table. She apologized, but the big man there was too busy studying his papers to notice.

'I hate the idea that nothing is left to chance', she continued furiously as she sat down. 'That means Max had to find Sidney at the flea market! It means he couldn't have not found him. Are you saying the coffee had to spill on my jeans? Stupid me, there I was trying not to spill it. Let's go – I need to change these jeans.'

They left the café, and Lea apologized again as she passed, but the big man just grunted. He leafed through the photos – hundreds of them. School photos: the boys in front sitting, those at the back standing. He had requested photographs from all the schools and now he was inundated with them. The Short Stocky Guy had been thumbing through the photos for hours and was beginning to feel as though he was about to go mad. He nodded to the waiter to bring him another espresso. He couldn't tell the difference between these kids – all he could tell was that none of them looked like the kid at the flea market. The Boss had phoned earlier that morning. He was beginning to get impatient. The man sighed and went back to the photos.

Meanwhile, Mr Ruche was in the Rainforest Library planning a journey north. On πR Fermat's weathervane, it pointed to number theory:

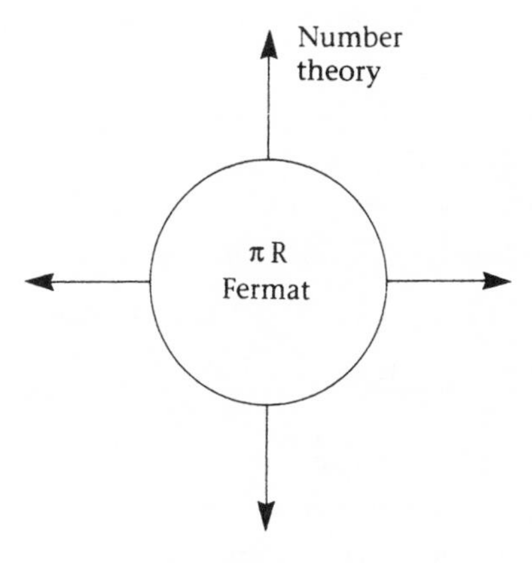

He took out the index cards related to number theory from Fermat's Complete Works, which were tucked at the back of the pile and began to read:

> In mathematics, the 'good' problems are those that can be simply stated...but whose solutions are fiendishly difficult. The greater the disparity between the simplicity of the formulation and the complexity of the solution, the 'better' the problem. In this sense, number theory is full of good problems.
>
> Fermat is unquestionably the master of number theory. Neither Pascal

nor Descartes, nor even contemporary mathematicians, can compete with him.

Number theory concerns the properties of numbers in themselves. From the division of numbers into odd and even, into prime numbers and non-prime numbers, the game consists of representing a number as a sum of squares or of cubes. How many squares, how many cubes?

N.B. Prime numbers have been an important part of cryptography for some time. Most modern codes are based on the properties of prime numbers.

Mr Ruche jumped when he read this. It was a clear indication from Grosrouvre that he had used codes. He would have to refresh his memory. In his notebook, he went back to what he had written about prime numbers:

A number is said to be prime if it has no factors other than itself and 1. With the exception of 2, all prime numbers are odd: 3, 5, 7, 11, 13, 17, 19, 23,...

All whole numbers can be divided into a unique set of prime factors.

If a prime number divides into the product *ab*, it will divide either *a* or *b* (i.e. a prime number cannot divide a product unless it can divide one of the factors of that product: one divisibility leads to another).

He wondered whether these were the properties of prime numbers Grosrouvre was referring to when he talked about codes.

He was distracted by the noise of Sidney fluttering insistently at the window. Mr Ruche wheeled himself to the door and let him in, and Sidney immediately settled on his perch.

Mr Ruche didn't know how to tackle the question of a code, and decided to continue reading the card. Grosrouvre listed a number of Fermat's results, preceded by a quote from the author:

The following is a summary of my dreams about numbers. I write it only because I realize that I shall not have the time to write out longhand all of my methods and proofs; in any case, this will point others in the right direction.

Every whole number is either a square, or the sum of two, three or four squares. More generally, every whole number is the sum of three triangular numbers, of four squares, of five pentagonals, etc.

A little further on, Grosrouvre quoted the famous 'Theorem of two squares'.

> All prime numbers (except 2) can be separated into two groups:
>
> > the first: 5, 13, 17, 29,...composed of numbers which, divided by 4, leave a remainder of 1 (which can be written as $4k + 1$);
> > the second: 3, 7, 11, 19, 23,...composed of numbers which, divided by 4, leave a remainder of 3 (written as $4k + 3$).
>
> All the numbers in the first group can be expressed as the sum of two squares and this can only be done in a single fashion; none of the second group may be expressed thus.
>
> Example: if $k = 3$, $4 \times 3+1 = 13$, a prime number, and $13 = 2^2 + 3^2$.
>
> Fermat then demonstrates his 'little theorem': If a is not divisible by p and if p is a prime number, then $(a^{p-1} - 1)$ is divisible by p.
>
> For this remarkable harvest of results, Fermat owes much to the *theory of infinite descent*. To prove that a problem cannot have a solution in whole numbers, Fermat proves that, if there were a solution, there would be another with smaller numbers.

Mr Ruche thought 'the theory of infinite descent' a rather beautiful name. 'But', thought Mr Ruche, 'how does that prove there is no solution?' He thought for a minute and then realized. 'Because there are always a finite number of whole numbers smaller than any given number, so there cannot always be a solution with smaller numbers. It works precisely because the descent can't be infinite. The theorem states that you have to keep going down, therefore there is a contradiction. The hypothesis must therefore be false, therefore there is no number which possesses the properties in question. QED.'

Mr Ruche smiled, rather impressed by his proof. He picked up Grosrouvre's next card and saw that it was headed 'The Birth of Fermat's Theorem'. At last! Mr Ruche decided he would be better not to go it alone. It was time to call a general meeting. However, curiosity got the better of him and he began to read:

> A friend of Fermat's, Claude Bachet de Méziriac, who had translated and published the six volumes of Diophantus' *Arithmetic*, gave him a copy as a gift. For Fermat, it was love at first sight, and he became obsessed by the problems posed by one of the great mathematicians of Alexandria.

Diophantine equations take the form

$$P(x, t, z) = 0$$

where P is a polynomial of multiple variables whose coefficients are whole or rational numbers (irrationals are out!). The difficulty is in the restrictions.

Although there are an infinite number of them, whole numbers make up only a tiny fraction of all numbers. The more restricted the area of research is, the less likely it is that one will find a solution.

Fermat annotated the pages of the *Arithmetic* – jotting down observations and adding solutions...but no proofs.

'Great!' muttered Mr Ruche. 'Here we go again – people scribbling all over books. I'm sure it didn't bother Grosrouvre. He's only happy when he's scribbling something in the margins of four-hundred-year-old books.'

Mr Ruche realized he was thinking about his friend in the present tense. For a while now, it was as though Elgar was everywhere: living by his side, planning every day for him. As long as the poet still sings of the hero, he still lives. According to the Greeks, it is only when the song ends that death really overtakes him. Looking at it from this point of view, Grosrouvre was more alive now than he had been for fifty years. Mr Ruche read on:

Two days after pleading a cause, Fermat died, without knowing whether he had won or lost the case. Some time earlier, fearing that his discoveries would be lost, he had asked his mathematician friends to collect them together – mostly from his letters – so that they could be published. Some of them began to do so, but the task was so great that they stopped. His son Samuel took up the task, and published almost everything that his father had ever written.

To the greatest work on number theory ever published, Samuel decided to add the annotations his father had scribbled in the margins of Bachet's translation of Diophantus. In Book II, opposite Problem 8, 'Divide a given square into two square numbers', Fermat had written this in the margin:

It is impossible to separate a cube into two cubes, or a fourth power into two fourth powers, or in general, any power except a square into two powers with the same exponent.

And he added, in the margin,

> I have discovered a truly marvellous proof of this, which, however, the margin is not large enough to contain.

That evening, in the Rainforest Library, Mr Ruche presented what he had learned to a general meeting of the Liard clan. 'If he hadn't spent his time ruining the book writing in the margins, he would have had more than enough room', said Mr Ruche. 'If he'd had a clean sheet of paper he could have written out his proof with room to spare.'

'If he'd done that, there would have been no story', said Jonathan.

'Yeah, and your friend Grosrouvre would have been pretty bored in Manaus', added Lea. 'I propose the motion: This house believes it was a good thing the margin of Bachet de...Bachet de who?'

'De Méziriac', Mr Ruche reminded her.

'This house believes it was a good thing the margin of Bachet de Méziriac's book was not large enough. I demand a vote!'

Though Sidney could not vote, the motion was carried unanimously.

'It was thanks to the narrowness of the margin', Perrette concluded, 'that your friend was able to solve Fermat's Last Theorem.'

'I think it would be better to say that he thought he'd solved Fermat's Last Theorem', said Jonathan. 'Just because he said in his letter that he had proved it doesn't mean that he had. It only proves that he thought he'd solved it.'

Perrette looked at him quizzically: 'What about you? Do you think he solved it or not?'

Everyone was silent, and all eyes turned to Jonathan. He looked at Perrette. 'No', he said, 'I don't.'

Mr Ruche opened his mouth, but no words came out. Then, with difficulty, he said, 'But why, Jonathan? Why would you not want him to solve it?'

'Well I think he did solve it', said Perrette, coldly, 'at least I hope he did.'

There was an uncomfortable silence, broken by Jonathan: 'Well, whether you like it or not, it was the fact that he kept it secret that killed him.'

Mr Ruche was speechless.

'But', Max spoke up, 'if Grosrouvre hadn't kept his proof secret, there wouldn't be any story – and anyway, sometimes it's better not to know everything.'

Max hadn't missed anything in the heated exchange. As always happened when a discussion became serious, all his senses opened up to what was going on. 'Anyway, you have to die of something', he said, his eyes shining, 'and Grosrouvre died of maths. It's probably the best thing that could have happened to him.'

'Could we get back to Fermat?' suggested Perrette. 'If I understand it, Fermat's theorem is telling us something that can't be done.'

'Exactly', said Mr Ruche.

'Can we write it out so we can understand it this time?' asked Lea.

'Yes, I'd like to see if I can understand it', said Perrette.

Lea wrote it out on a piece of paper and drew a box around it:

> *It is impossible* to find four whole numbers x, y, z and n, where x, y, z are not equal to 0 and n is greater than 2, such that
>
> $$x^n + y^n = z^n$$

'It would be easier to say that it's impossible to divide a number to the nth power into the sum of two numbers of the same power except for squares', said Jonathan. 'See, it's easy.'

'Well, go on then, prove it!' Lea challenged.

'I meant it's easy to say', said Jonathan. 'I don't think we should try to prove it.'

'In any case', said Perrette, 'dinner is nearly ready.'

The twins, who were ravenous, raced out, followed by Max. Mr Ruche stayed behind to tidy up.

'Don't be too long, Mr Ruche', called Perrette.

As he filed away his notes and Grosrouvre's index cards, he spotted a quotation on a piece of paper. It was an epitaph for Diophantus written by Metrodorus. Mr Ruche read it, and became more puzzled:

Traveller, here lies Diophantus.

O, great prodigy, science itself will give you the measure of his life. God granted that his boyhood lasted one-sixth of his life; his beard grew after one-twelfth more; he married after one-seventh more; and his son was born of that marriage 5 years later.

Alas, his poor son felt the cold of death when he had lived to just half his father's age. Four years later, his father found consolation for his affliction when he reached the end of his life's span. What was the span of his life?

The back of the card was blank, though Mr Ruche knew that Grosrouvre would not have written the answer. 'Well then', thought Mr Ruche, 'we'll see if all this maths study has left me fit to solve this.' He set to work, trying to remain as rigorous as possible.

'It's an equation with one unknown. Al-Khwārizmī said the first thing to do was name the thing you seek – how long Diophantus lived. We'll call it v.'

'Now, what do we know about v? We know that it is divided into sections which, added together, equal v.' Mr Ruche jotted some figures on the back of the card:

His boyhood lasted one-sixth of his life:	$\frac{v}{6}$
His beard grew after one-twelfth more:	$\frac{v}{12}$
A seventh more before he was married:	$\frac{v}{7}$
And five years later, his son is born:	$+5$
He has lived half his life before his son dies:	$\frac{v}{2}$
And four years later, he dies:	$+4$

Mr Ruche set it out as an equation:

$$v = \frac{v}{6} + \frac{v}{12} + \frac{v}{7} + 5 + \frac{v}{2} + 4$$

As he was writing, the studio door opened and Max came in with Sidney on his shoulder. Max couldn't help noticing that Mr Ruche was puzzling over something, and asked what it was.

'I'm trying to work something out.'

'What is it? Can I have a look?'

'Go ahead, look all you want.'

Max leaned over Mr Ruche's shoulder and smiled. 'What does v stand for?'

'Someone's life.'

'It's positive, then. I mean, it's a positive number because a life that was a negative number would be very strange. Anyway, I'd better go.'

'Max, no! I need you to help me!'

'Mum said to tell you dinner's ready', Max said and glanced at the paper again. 'It's easy, Mr Ruche: find the common denominator, simplify the fractions – pretty standard stuff.'

'Pretty standard stuff for a twelve-year-old', thought Mr Ruche as the door slammed behind Max. He simplified the equation:

$$v = \frac{75v}{84} + 9 = \frac{25v}{28} + 9$$

$$\text{Therefore} \quad v - \frac{25v}{28} = 9$$

$$\text{Therefore} \quad \frac{28v}{28} - \frac{25v}{28} = 9$$

$$\text{Therefore} \quad \frac{3v}{28} = 9$$

$$\text{Therefore} \quad v = 28 \times \frac{9}{3}$$

He started calculating, and then screwed the paper into a ball. 'Oh, I give up!' But he picked it up again and unfolded it, and the solution was right there in front of him:

$$v = 28 \times \frac{9}{3} = 84$$

Mr Ruche sat back in amazement – Diophantus had died aged 84, just like al-Khayyām and Grosrouvre.

'Mr Ruche!' Perrette called from the balcony opposite. 'Your soup is getting cold!' Mr Ruche smiled, stuffed the paper into his jacket pocket and left the studio.

'I've found something for you', Lea said to Mr Ruche after dinner. Mr Ruche looked up, curious to know what it was.

Lea was signalling to Sidney, who piped up on cue, 'There are three principal objectives in seeking the truth: the first, to find it when one seeks it; the second, to prove it once one has found it; and the last, to distinguish it from falsehood when it is examined.'

Lea went on: 'The first principle concerns us and what we are doing here – finding the truth we are searching for; the second is about Grosrouvre. By working on a theorem he is trying to prove a truth he already has. I hope he fared better than we have!'

Mr Ruche sat up. 'Pascal – from *De l'esprit de géométrie et de l'art de persuader.*'

'Bravo!' Jon-and-Lea and Perrette were genuinely impressed. Mr Ruche looked modest. 'Culture is what you remember when you've forgotten everything else. I could have been…' He stretched out his arms. They all looked at him. He let his arms fall back onto his knees. 'I could have been a contender', he smiled.

'We wouldn't want you any other way', said Max.

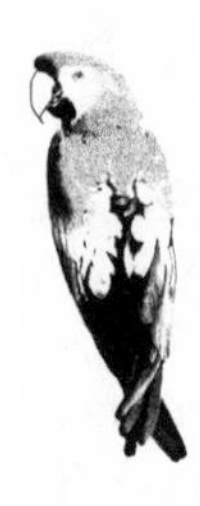

CHAPTER 20

A Handful of Feathers

π in the sky; Euler and the magical number e; John Napier and logarithms & a fit of apoplexy

Mr Ruche was woken in the late morning by a clattering coming from the apartment. Though he was still half asleep, he knew it couldn't be Perrette as she spent Monday mornings touring bookshops in the Latin Quarter. Max and the twins were at school, so who...? Something crashed to the floor, then he heard Sidney squawking. There was a silence for a minute, then the sound of running footsteps. Then nothing.

'Sidney!' called Mr Ruche as he wheeled himself quickly to the lift. The apartment door was wide open. He pushed the wheelchair as fast as he could. There was a horrible, chemical smell wafting from the apartment. He put a handkerchief over his mouth and went in. The perch was lying on the floor. Seed and water were everywhere, and a handful of feathers were strewn nearby. Sidney was gone – he had been kidnapped!

Mr Ruche nervously wheeled himself round in circles as he waited for Perrette to come back. He tried to tidy up but found it difficult. He was trying to set the perch upright when he heard her on the stairs. She guessed something was wrong as soon as she saw that Mr Ruche's door was open. She rushed up the stairs and could see at once what had happened. Mr Ruche looked terrible – slumped in his chair, his face white. He had become very attached to Sidney while they worked on the sessions in the studio. Mr Ruche had never met a creature as intelligent as Sidney before.

Perrette recognized the chemical smell immediately – it was chloroform. Sidney had been knocked out, but the feathers on the floor proved that he had put up a fight. Perrette picked them up carefully and put them on the table. She stood the perch up, then got a bucket and cleaned up the food and the water on the floor. She checked the room, but nothing else had been taken. Whoever it was had come for the parrot.

Mr Ruche asked her to take the feathers off the table.

'You don't want me to throw them out, do you? They might be clues. We have to keep them for the police.'

'Yes, but Max will be back from school soon and he shouldn't see them', said Mr Ruche.

'It's a good job you didn't get here while they were still here', said Perrette. 'These guys knew what they wanted and they wouldn't have stopped at anything.'

They heard Max's footsteps on the stairs. Mr Ruche just had time to gasp 'Perrette, the feathers!' She picked them up and put them in her pocket just as Max ran into the room. He saw them both looking at the empty perch and immediately asked, 'Where's Sidney?'

Perrette told him everything. 'The bastards', his dark eyes glittered angrily, 'they better not have hurt him or...' His voice was so threatening that even Perrette was a little afraid of her son.

'I know who did this', Max muttered to himself.

'Who?'

'The gang from the flea market. They're trafficking in animals.'

'What gang?'

'The guys from the flea market, Mum, the ones who had Sidney in the first place.'

'But he's been here for months. How could they find him?'

Max told them about his trip to the pet shop and how the salesgirl had behaved strangely.

'But if they followed you here, why did they wait so long before doing anything?' Perrette asked, upset.

'I'm sure they didn't follow me', said Max, 'I was really careful.'

'Then how did they find out where you lived? It's the only way...'

'I wasn't followed, I told you!' He seemed very sure of himself.

'I'll call the police', said Perrette.

'No, Mum, you can't! If we call the police, we'll be in trouble. If they find Sidney, they'll take him away from us. It said so on the poster – any bird brought into the country illegally will be confiscated. If we found him and they took him away again I couldn't bear it.'

He told them what he had learned at the pet shop about the papers you needed: medical certificate, proof of purchase, proof of quarantine and vaccination.

'What do you think, Mr Ruche?'

'Max is right. The first thing we should do is go to the pet shop. We have to find that salesgirl.'

'I'll go tomorrow.'

'I think maybe we should go now', said Mr Ruche.

'I can't leave the bookshop closed all day', said Perrette. She hesitated a minute, then said, 'OK, you're right. I'll put up a notice.'

'You do that', said Mr Ruche, 'put up a sign saying "Bookshop closed owing to kidnap of parrot".'

'OK, OK – I'll just close up the shop. I won't leave a notice.'

'There's no need to close it at all – you go and I'll look after the shop.'

'But you haven't...it's been ten years...'

'You think I won't remember? I did run the place for thirty-five years.'

Perrette refused to let Max come with her. She walked around the pet shop but couldn't see the girl he had described, so she asked to see the manager. He was not very helpful. Perrette described the girl. 'Oh yes, yes. Giulietta – Giulietta Giletti. I'm afraid she doesn't work here any more. She left last week. She was only here for a month or so. I wanted her to stay, but she said she had other plans. Are you a friend or relative?'

The manager refused to give Perrette Giulietta's address, so she had to explain what had happened. She told him about Max's visit to the pet shop and how the girl had behaved strangely, but she didn't tell him that Sidney had been kidnapped. She said she thought that Giulietta was involved in trafficking animals.

'Madame, are you insinuating that this establishment...'

'Of course not, of course not...'

'I'm horrified you could even suggest such a thing. This is a well-known and highly respected shop which has been here for more than a century. You should know that all pet shops are regularly checked, which is more than I can say for other businesses. The authorities are very strict. They regularly check the paperwork on our birds and ensure they have all been vaccinated.'

He calmed down a little and continued, 'It is true that a black market in exotic birds has sprung up in Paris over the last couple of years. It has done a lot of harm, but everyone knows where it goes on – at the flea markets.'

It all made sense, thought Perrette – Max was right all along.

The manager invited her into his office. He took out a folder of newspaper cuttings. The first related to a police operation code-named 'Birdy', the second to operation Romeo, in which five traffickers were apprehended and 499 birds were retrieved: cockatoos and macaws mostly, but no parrots.

The manager looked in his card index and handed a card to Perrette. 'This is the address you were looking for.'

Perrette took the card and thanked the manager. She checked out the address immediately, but there was no one there by the name of Giulietta Giletti. This simply confirmed her suspicion that Max had been followed back to the rue Ravignan.

Max locked himself in his room all day. He was angry at himself for not being there to help Sidney, but he knew he wouldn't be allowed to take a parrot to school with him. They had guide dogs for the blind, he thought, why not parrots for the deaf? He came downstairs for the general meeting with the household that evening. They decided that it would be a mistake to give up their work. Sidney had been one of the main players in their research, but it had to go on.

Early the following morning, Mr Ruche went to the Rainforest Library, though his heart was no longer really in it. The next name on Grosrouvre's list was Leonhard (with an 'h') Euler, born in Basle in 1707. Mr Ruche had never heard of Euler, so he decided to cheat and looked him up in a mathematical dictionary. There he was, just after Euclid. There were eight pages dedicated to him. Mr Ruche began making notes:

> Euler grew up near Basle, Switzerland, and studied at an early age under Jean Bernoulli. He finished studies at the University of Basle when only

fifteen years old. From 1727 to 1741, Euler worked in St Petersburg, Russia, and then moved to the *Akademie* in Berlin. In 1766 he returned to St Petersburg, where he remained.

Euler made great strides in modern analytic geometry and trigonometry. He made decisive and formative contributions to geometry, calculus and number theory. In number theory he did much work in correspondence with Christian Goldbach. He integrated Leibniz's differential calculus and Newton's method of fluxions into mathematical analysis. In number theory he stated the prime number theorem and the law of biquadratic reciprocity. He also introduced beta and gamma functions, integrating factors for differential equations, etc.

He was 'far and away the most prolific writer in the history of the subject' writes Howard Eves in *An Introduction to the History of Mathematics*. His complete works contain 886 books and papers.

Around the world, notations conventionalized by him or made in his honour are read, written and spoken thousands of times every day:

e for the base of the natural logarithm (a.k.a. 'the calculus number').

a, b, c for the side lengths of a triangle ABC.

$f(x)$ for a functional value.

R and r for the circumradius and inradius of a triangle.

$\sin x$ and $\cos x$ for values of the sine and cosine functions.

i for the imaginary unit, the 'square root of -1'.

Σ, capital sigma, for summation.

Δ, capital delta, for finite difference.

Mr Ruche was disturbed to discover the list of all of Euler's contributions: Euler's angles, Euler characteristic, Euler circle, Euler circuit, Euler–Mascheroni constant, Euler line, Eulerian numbers, Euler triangle, Euler's first integral (beta function), Euler's second integral (gamma function), Euler polynomials, Euler's phi function, Euler's identity (zeta function). Euler force (critical load), Euler's formula ($e^{i\theta} = \cos \theta + i \sin \theta$), Euler–Maclaurin summation formula, Euler's equation (in the calculus of variations), Euler's equation (in differential equations), Euler's equation of motion of an ideal fluid, Euler's equations for the rotation of a rigid body, Euler's equation on normal curvature, Euler method and improved Euler method for differential equations, Euler's addition theorem for elliptic integrals, Euler's criterion for quadratic residues, Fermat–Euler theorem, Euler–Lagrange theorem (every positive integer is a sum of at most four squares), Euler's officer problem, Euler's theorems on partitions, Euler's theorem for four collinear points, Euler's theorem for polyhedra ($V - E + F = 2$), Euler's theorem for primes, Euler's theorem for rotation of a coordinate system, Euler's theorem for a triangle ($d^2 = R(R - 2r)$, where d is the distance between the circumcentre and incentre).

Mr Ruche didn't recognize all these terms. In fact, he was unclear about many of the concepts, but it was reassuring to find some he had recently discovered: complex numbers, circumscribing circles, algebraic equations, polyhedra, differential equations. If nothing else, the last six months had improved his vocabulary.

Impressed by the extent of Euler's work, Mr Ruche wheeled himself over to Section 3 of the Rainforest Library, where Euler was next to Descartes. He wheeled himself on and on down the shelves: 75 volumes of mathematical ideas – his letters were not included – written by one man. Euler was almost a library in his own right.

Mr Ruche felt suddenly overwhelmed. How could he be expected to tackle a body of work such as this. 'I can't do this at my age', he thought, but that made him shudder. 'This is the only age I have to do it in.'

He went back to work – but where should he start? He opened a volume at random and his eye fell on an equation in the middle of the page:

$$\frac{\pi^2}{6} = 1 + \frac{1}{4} + \frac{1}{9} + \frac{1}{16} + \cdots + \frac{1}{n^2} + \cdots$$

Mr Ruche tried reading it aloud: 'one sixth of pi squared is equal to the sum of...the inverse...of a series of whole numbers squared'. He was quite proud that he had managed it. Reading the equation aloud was like decoding it – bringing it out into the light of day. π squared...that was it. It was time to tackle π, and Mr Ruche knew just where to go. Besides, he thought, it would do Max good to get out a bit.

They took the métro to the Place de la Concorde, and Max pushed Mr Ruche's wheelchair up the Champs-Élysées towards the Arc de Triomphe. Halfway up they came to the Grand Palais, with its great, vaulted glass roof.

The palace had been built for the Great Exhibition of 1900, Mr Ruche explained to Max, but because the ground sloped down to the river, it had to be levelled. Instead of levelling it with earth, the builder had buried thousands of oak posts. Eighty years later, the Grand Palais started tilting towards the Seine.

'I first came here in 1937', said Mr Ruche, 'the year the government passed a law saying everyone was entitled to four weeks' holiday – though we didn't call them "holidays" in those days – we called them "days off".'

Mr Ruche wheeled himself across the mosaic floor of the great elliptical entrance hall to the Palace of Discovery. 'I remember it like it was yesterday. Lectures were finished for the year, and one morning Grosrouvre practically dragged me here. We'd only just walked in the door when he pointed out that the hall wasn't circular.'

Max threw his head back and looked up at the great glass dome and the light pouring through it. When they came to the centre of the ellipse, Mr Ruche turned right towards the stairs.

'We took the stairs two by two because Grosrouvre was in such a hurry.' Mr Ruche frowned and stopped his wheelchair at the bottom of the vast stairway. There was no lift, no wheelchair access. Disappointed, they were about to turn round and leave when a group of schoolboys, who had been watching as they waited for their teacher to buy tickets, came over, picked up the wheelchair and carried it up the stairs. Mr Ruche was bounced this way and that, laughing all the way. Max ran up the steps behind them. They arrived on the first floor in no time. The boys weren't even out of breath.

Mr Ruche took control of the wheelchair again, and the procession took on a more dignified air. Mr Ruche remembered something that Grosrouvre had recited to him on their visit in 1937: 'Science rose from the oceans of the Abstract, wrapped in the robes of Aphrodite, blown onward by the spray. A garland fluttered under a dome borrowed from a cubist film and on it were the first 700 decimal places of the great number π.'

They had arrived at the Temple of π. It was a great, round hall, and a band running round the room bore the names of famous mathematicians. High above was a spiral frieze, which wound around the room several times. On it, in groups of ten, were the first 707 decimal places of π.

Max was overwhelmed by all this mathematical graffiti. He started at the 3, jumped the decimal point and read 1415926535, written in red, 8979323846 in black, 2643383279 in red. He speeded up: red, black, red, black, his eyes leaping from number to number. Max was spinning round now. He was dizzy, he had never seen so many numbers in his life. He spun around for the fourth time. He felt like he was going to take off – he whizzed past the last number, unable to stop. When he finally managed to tear his gaze away from the frieze, he had numbers dancing before his eyes and the room was swimming. He held on tightly to Mr Ruche's wheelchair.

The room fell silent and a guide came in. He stood beside a flip chart and began immediately: 'On a plane, a line is the shortest distance between two points. If you feel like a walk, you can take a circular route. It is the longer way round. How much longer?' He drew a diagram on the flip chart. 'It will be exactly $\frac{1}{2}\pi$ times longer.'

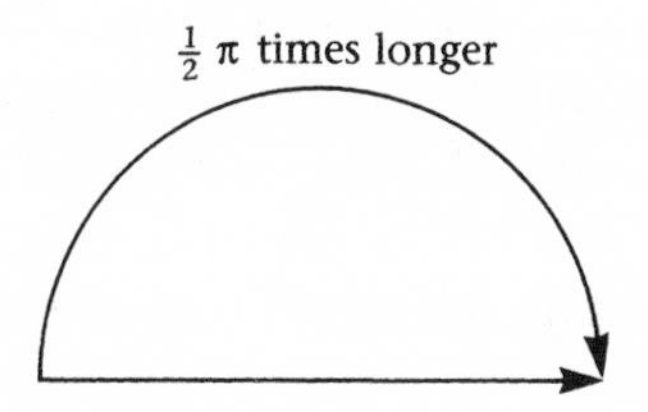

'π is the relationship between the line and the circle. It has a very long history – from Babylon, to Ahmes in Egypt, Archimedes in Greece; Āryabhata in India, Zu Chongshi in China...Al-Kāshī in Samarkand took it to 14 decimal places. Ludolph van Ceulen had 35 decimal places of π engraved on his tombstone...'

The guide wrote something on the flipchart. 'Now we come to formulas', he announced. 'François Viète devised an astonishing formula for calculating π which uses only one other number, 2. It is based on a series of square roots. It was the first infinite formula:

$$\pi = 2 \times \frac{2}{\sqrt{2}} \times \frac{2}{\sqrt{2+\sqrt{2}}} \times \cdots$$

'Everything depends on the denominators, which get bigger and bigger – if they didn't, the product would be infinite.'

In the seventeenth century, he went on to explain, calculating π became an obsession in Britain. The different formulas proposed involved sums, products, quotients. The first of these was by John Wallis. As he wrote down the formula, the guide explained it to his audience:

'The numerators are pairs of whole, even numbers: 2 times 2 times 4 times 4 times 6 times 6, etc. The denominators are whole, odd numbers, also in pairs: 3 times 3 times 5 times 5 times 7 times 7, etc.:

$$\frac{\pi}{2} = \frac{2\times 2\times 4\times 4\times 6\times 6\times\cdots}{3\times 3\times 5\times 5\times 7\times 7\times\cdots}$$

'Then came William Brouncker, the first president of the Royal Society. He devised a fraction for calculating π which is very different from those we normally use. It is a continuous fraction. The numerator is a whole number added to a fraction, and the denominator is also a whole number added to a fraction, whose denominator is also a whole number added to a fraction, and so on. This definition is by Leonhard Euler and uses the squares of odd numbers.'

He wrote out the formula, bending lower and lower as he went on.

$$\frac{4}{\pi} = 1 + \cfrac{1}{2 + \cfrac{3^2}{2 + \cfrac{5^2}{2 + \cfrac{7^2}{\cdots}}}}$$

'It's the *Titanic*!' shouted someone. 'It's sinking!'

One of the boys who had helped carry Mr Ruche's wheelchair up the stairs shouted, 'Abandon ship, lads! We'll never be able to read it!'

'Go on, Henry, dive!'

Henry took a deep breath, and the other boys watched as he held his breath for a minute, then someone shouted 'Go!'

Calmly, he let his breath out and as he did read the formula: '4 over pi equals 1 plus 1 over 2 plus 3 squared over 2 plus 5 squared over 2 plus 7 squared...'

He made it to 27 before his breath gave out.

The guide smiled and waited as the audience applauded before continuing: 'Next came Isaac Newton, James Gregory and John Machin. Newton wrote to one of his friends, "I have nothing to do at the moment, so I have just calculated π to 16 decimal places." John Machin was the first to calculate it to 100 places. At the end of the seventeenth century, Gottfried Leibniz devised an infinite sum which also used a series of odd numbers:

$$\frac{\pi}{8} = \frac{1}{1 \times 3} + \frac{1}{5 \times 7} + \frac{1}{9 \times 11} + \cdots$$

'Although these are all pretty formulas, they're not necessarily "good" formulas, in that they don't elegantly generate decimals. Some of them converge very slowly, others move more quickly. Then we come back to Leonhard Euler.'

Under his breath, Mr Ruche recited, 'The sum of the inverse of a series of whole numbers squared.'

But the formula the guide wrote on the flip chart was different to the one in Mr Ruche's notebook:

$$\frac{\pi^2}{6} = \sum_{n=1}^{\infty} \frac{1}{n^2}$$

'I can see that some of you are puzzled', said the guide. 'The symbol after the equals sign is the Greek capital letter sigma, the equivalent of S in the modern alphabet. It's simply a shorthand way of representing a sum, usually an infinite sum. It reads "the sum from $n = 1$ to infinity". The symbol at the top – the eight on its side – represents infinity, the number towards which n tends.'

'This was the beginning of the great decimal race. The number of decimal places quickly leapt to 127, then to 140. Professional calculators began to enter the race – "decimal chasers", they were called. 200 decimal places were reached in 1844, and then there was a huge leap to 440 by William Rutherford in 1872, who felt his record would be safe for some time. He was wrong – two years later, William Shanks announced that he had reckoned π to 707 decimal places. He was feted as a hero, and deservedly so, as he had spent twenty years calculating his result.'

Mr Ruche tried to imagine what William Shanks's life must have been like. Sitting at his desk every morning for twenty years and saying, 'Now, where was I?' He felt a little sick at the thought.

The numbers on the frieze were William Shanks's, the guide explained, whose record remained unbroken for seventy-one years.

'The Second World War had barely ended when, in 1947, a man named Ferguson discovered...' The guide stopped, leaving his audience in suspense. He took a long ruler and pointed to the '9' in the fourth turn of the spiral '...that the 528th decimal was wrong!'

The crowd let out a long 'Aaaah!' which echoed strangely around the hall.

'Well, well!' thought Mr Ruche. 'Grosrouvre spent all that time staring at the wrong numbers.' He started to laugh, he couldn't stop himself. 'To think', thought Mr Ruche, 'Grosrouvre never realized, because by the time Ferguson worked out it was wrong he was in the Amazon.'

One of the boys interrupted, 'But if the 528th figure is wrong, then the ones that follow it must be wrong too.'

'Exactly!' said the guide.

'But that means that 180 of the decimal places up there are wrong!'

'They were – but in 1949 the management had the last 180 places following the "9" erased and replaced with the correct numbers.'

The audience stared hard at the numbers, but nothing in the colour or the shape of the figures gave anything away. The frieze looked as though it had never been altered.

The guide picked up his story again. 'It was also in 1949 that the next barrier was broken – 1000 decimal places. After that, the calculations were done using machines. Early computers were programmed to calculate π to ever more decimal places. They reached 10,000 places in 1958; 100,000 in 1961, 1 million in 1973, 10 million in 1983, 100 million in 1987 and 1 billion decimal places in 1989!'

'Two or three little things before we finish', said the guide. 'You shouldn't think that π exists only in the world of pure maths. It can also be found, here and there, in physical phenomena and in cosmology.' He pointed to the dome above and pressed a button. The lights went out, and a projection of the stars appeared.

'Some astronomers believe that π is present in the heavens. If every star in the sky is referenced by two coordinates expressed as whole numbers, the probability that those numbers will have no common factor is $\frac{6}{\pi^2}$.'

The lights in the dome came on again.

'Here on earth, π is linked to the great rivers of the world', the guide continued. 'Large rivers flow slowly and tend to meander. If one compares the length of the river from source to estuary as the crow flies with the true length of the river, measuring all the meandering, the ratio is almost exactly 3.14. The flatter the topography, the closer the ratio is to π. The Amazon is probably the best example.'

Mr Ruche heard Max say, quite seriously, 'There's pi in the sky and pi on earth.'

'As you leave the hall, you'll notice an equation inscribed above the doorway. This is Euler's equation, and perhaps the most beautiful in all of mathematics.'

As they filed out, everyone looked up and read:

$$e^{i\pi} = -1$$

Mr Ruche craned his neck to look up. It was beautifully simple, true – but was it beautiful? He studied it. Five symbols: π; the 'equals' symbol, which Recorde had invented; '–1', which still reminded Mr Ruche of descending a multi-storey car park; 'i' for imaginary', which Euler invented himself. Then there was the 'e', which he had seen for the first time that morning.

'Do you think it's beautiful?' he asked Max, who was also staring at the equation.

'I dunno...beauty is in the eye of the beholder.'

Suddenly, they heard a voice: 'You must be really interested in that equation.' It was Perrette, with Jon-and-Lea. 'We've been standing here for five minutes.'

Mr Ruche couldn't think of anything to say except, 'Do you know what e is?'

'Yes, we do', chorused Jon-and-Lea.

The boys who had helped Mr Ruche up the stairs were no longer there to carry him down again, but it didn't matter because Mr Ruche had all his family with him now. Jonathan and Max took one side, Perrette and Lea took the other.

The people they passed on the way down could not help admiring Mr Ruche, who looked like a king being carried in a procession. He tried to look impassive, but secretly he was proud. Suddenly he remembered that it was still the middle of the afternoon.

'Perrette, you've closed up the shop!'

'Yes, Mr Ruche, I did what you suggested. I took the day off.'

Despite their boasts, neither half of Jon-and-Lea knew the first thing about e, but that evening they determined to discover what it represented. They were surprised to find that it was simply a number, like 1, 2 or π, but – like π – it could not easily be expressed in decimals. Not only did e have an infinite number of decimal places, but it was irregular – there was no way of knowing what the next digit would be without calculating it. Lea had her own definition:

'A number that goes on for ever and, worse still, seems to do it any old way.'

$$e = 2.718\,281\,828 \cdots$$

As usual, they shared the work, which meant that Lea did everything and Jonathan did nothing.

'e is a very interesting number', Lea explained to him later. 'Listen, let's suppose you'd earned enough money to pay for our trip to Manaus. Let's call it *C* for cash.' She wrote the letter on a sheet of paper.

'In the meantime, you decide to put it in the bank. You're really lucky, and you find a bank that will pay 100 per cent interest. After a year, you would have

$$C + C = 2C$$

'You'd have doubled your cash.

'Now, if your interest was paid every six months instead of every year and you reinvested all your money, after a year you would have

$$C(1 + \tfrac{1}{2})^2 = 2.25C$$

'So you would have more than doubled your cash.

'Right, suppose your interest was paid quarterly and you reinvested everything. After a year, you'd have

$$C(1 + \tfrac{1}{4})^4 = 2.441C$$

'You'd have made even more money. Now, if interest was paid monthly, you'd get

$$C(1 + \tfrac{1}{12})^{12} = 2.5996C$$

'More again! Suppose it was paid daily:

$$C(1 + \tfrac{1}{365})^{365}$$

'If it was every minute or every second, you'd get even more. You'd be raking it in, wouldn't you? Your money should be a hundred or even a thousand times more. You'd think about sharing the money with me 'cos you're earning so much interest. But, you're not! Your compound interest might be paid and compounded every nanosecond, but you'd never even triple your money: the most you would get is 2.718281828 – you'd never be more than e times richer than you were at the start.'

'I don't like this e business', said Jonathan. 'It sounds like some sordid banker's trick to keep them rich and us poor. It's not e at all, it's aaargh!'

'Don't throw out the baby with the bathwater. Exponentials are pretty amazing things. You can find them all over the place – in nature and in society, in the way plants grow, epidemics spread, in the way populations expand. "When the degree of development is proportional to the state of development, it becomes exponential." '

'So, the rich get richer and the sick get sicker...'

'It's worse than that! The richer you are, the richer you get and the faster you get richer. I don't know how to explain it. It just grows – and the more it grows the faster it grows. The best way to explain it is using maths. If something grows steadily, say at a rate of $2x$, then its growth is linear. It grows at a constant rate – the rate is equal to its derivative, which is 2. But if the growth is x^2 – a parabola – then it grows at a rate of...'

'...at a rate of $2x$', said Jonathan.

'But', said Lea, 'the growth of its growth is still constant and equals its derivative, 2.'

She saw Jonathan's puzzled face.

'Don't let me down, now – if I can follow it, you can follow it!'

'Oh no! You see, I'm Epiphanus and you're Hypatia, his talented sister.'

'Oh, don't be so dramatic! Anyway – if your growth is e^x, then, not only does it grow and not only does its rate of growth grow, but the growth of the rate of its growth grows, and so on...Why?'

Jonathan shrugged, so Lea answered: 'Because the derivative of e^x is e^x. It's an exceptional case – it's the only case where the derivative is equal to the number.'

The next morning, they talked to their maths teacher about e. He was impressed with their research, but surprised that they had forgotten that e was linked to logarithms. The two geniuses were embarrassed and decided they would not set foot in class C113 again until they had mastered logarithms. This time, Jonathan did all the work and Lea did nothing.

When he had finished, he presented his research to her, starting with a definition: 'If a, b and c are three numbers such that $a^b = c$, then b is the logarithm of c in base a:

$$a^b = c \Leftrightarrow b = \log_a c$$

Lea stared at him, horrified. He smiled and explained:

'OK: $10^2 = 100$, so the logarithm of 100 to the base 10 is 2:

$$\log_{10} 100 = 2$$

'$10^3 = 1000$, so the logarithm of 1000 to the base 10 is 3:

$$\log_{10} 1000 = 3\text{'}$$

Lea nodded, reassured.

'In base 2, for example, the logarithm of 8 is 3:

$$\log_2 8 = 3, \quad \text{since } 2^3 = 8$$

'Therefore, there are as many possible bases as there are numbers. However, 1 is excluded, as are negative numbers as the base for logarithms.'

'Why not all numbers?' asked Lea.

'Five minutes ago, you knew nothing about logs at all and now you want them to have bases in all numbers!'

'Well, it doesn't seem the same if you leave out 1.'

'You'll get used to it. There is no log with a negative base, or to base 1. It still leaves us with quite a few. All logs have a common point:

$$\log_a 1 = 0$$

'What about e?'

'The logarithm to base e is called a natural logarithm. It's usually written as $\log_e$, or ln. It's sometimes called a Napierian logarithm, after the guy who invented logarithms in 1614, John Napier.'

They went to the Rainforest Library together and found Napier's *Mirifici logarithmorum*. Jonathan read out the title page:

> *Mirifici logarithmorum*
> A description of the marvellous rules of logarithms and their uses in trigonometry and in all mathematical calculation. The most comprehensive and simple explanation, devoid of all complications

The first fifty-six pages presented definitions and explanations, and then there were tables and tables of logarithms. These were the first of the log tables which had been used for centuries in thousands of calculations. Since the invention of calculators, they had become merely a curiosity in second-hand bookshops.

They quickly discovered the 'marvellous rules' that Napier had mentioned. The efficiency of logarithms could be summed up in four simple rules:

The logarithm of a product is the sum of its logarithms

$$\log xy = \log x + \log y$$

The rest was easy: to perform a division, one simply subtracted:

$$\log x/y = \log x - \log y$$

To raise something to a power, one simply multiplied:

$$\log x^n = n \log x$$

And calculating roots required simple division. To calculate a square root, for example, you simply divided by two:

$$\log \sqrt{x} = \frac{1}{2} \log x$$

'This makes everything much easier', said Jonathan. 'To find out the 17th root of 1789:

$$\sqrt[17]{1789} = \log 1789 \div 17$$

'Then you just look up the result in the log tables, and you've got it. Logs must have been pretty revolutionary – I mean, you could spend days trying to work out square roots back then. Then, suddenly, Napier gives you some log tables and you can do it in a minute.'

'Nowadays, you'd just use a calculator', said Lea.

Pleased with their new-found knowledge, Jon-and-Lea went to see Mr Ruche, who was lying on his bed in his garage room. They talked for a long time. When they had finished, Mr Ruche said, 'That's all very well, but it doesn't explain why the formula in the

Temple of π is the most beautiful in the whole of mathematics.'

'Beauty is in the eye of the beholder', said Jonathan.

'That's what Max said', Mr Ruche smiled.

'Where is Max?'

'At the flea market – he's there every day now, looking for the guys who kidnapped Sidney. He's sure it's the same guys he saw there.'

'It could be dangerous', said Jonathan.

'Once he's made up his mind about something, there's no stopping him', said Lea, 'you should know that.' She sat at the end of Mr Ruche's bed and announced, 'The story of e, Part 2. John Napier spent twenty years compiling the log tables.'

'Another lunatic!' said Mr Ruche, putting a big cushion behind his head. 'I can't think of anything I would want to spend twenty years of my life doing.'

There was a knock at the door. It was Max, back from the flea market. He didn't expect to find Jon-and-Lea there, and went to leave.

'Max, hang on!' said Jonathan, moving across to his brother at the door. 'We promised you something, so here it is. It's about the equation Mr Ruche seems to obsessed with.'

'I am *not* obsessed', objected Mr Ruche, 'I've been told it's the most beautiful equation in the world and I'm simply trying to work out why.'

'Max agreed to find the answer for you, but since he's been busy, we've done the research for him', said Lea to Mr Ruche.

Max nodded in agreement, and took the paper Jonathan handed him and unfolded it. He read it aloud to Mr Ruche: '$e^{i\pi} = -1$, which can also be written as: $e^{i\pi} + 1 = 0$, is the simplest and therefore the most beautiful formula, in that it contains the fundamental numbers of mathematics: 1, 0, π, e and i.'

That afternoon, Mr Ruche went back to his biography of Euler and found to his shock another library in flames.

> Euler could smell burning...On a May afternoon in 1771, a fire spread quickly through St Petersburg. More than five hundred buildings were destroyed in the inferno. Euler was alone in his house, working in his study. Flames licked at the door and the air was thick with

smoke. Euler was, by then, nearly blind and could not find the door. He was trapped.

Suddenly, a man burst into the room, lifted Euler onto his back and hurled himself back into the flames. Outside, where an anxious crowd waited, Peter Grimm appeared out of the smoke and put Euler down. Miraculously, neither man had been injured. Euler explained where his manuscripts were kept, and a human chain formed to try and salvage them.

Many of the manuscripts were saved, but Euler lost everything he had been working on recently. His library, which was in his office, was completely destroyed.

Mr Ruche felt a little sad to think of all the books that had been destroyed in the stories he had read since Grosrouvre's first letter. He looked up fondly at the Rainforest Library – he knew that he was lucky the books had arrived safely. But if the library burned he would be to blame. There was no alarm, no lock, no smoke detector, not even a fire extinguisher. He left the Rainforest Library and headed for the bookshop. He had to do something quickly. He explained the situation to Perrette.

Perrette waited until he had calmed down and suggested they call a security company. They could say they wanted to install a fire alarm in the bookshop and at the same time have one installed in the Rainforest Library. They could tell the security company that it was a stockroom, and put dust sheets over the shelves so that no one could see the books while the alarm was being installed.

It would be expensive, but when Perrette suggested they sell one of the books from the Library to pay for the installation, Mr Ruche shook his head. He decided to dip into his savings. Perrette agreed to arrange everything.

Reassured, Mr Ruche went back to his study. If Euler was on Grosrouvre's list, he thought, it was because of the fire. But, on reflection, he was puzzled. Euler's story was exactly the opposite of what had happened to Grosrouvre: his house didn't burn, and neither did his manuscripts, but his library was destroyed. In any case, thought Mr Ruche, Grosrouvre's letter had been written a month before the fire in Manaus, so it couldn't have been on his mind when he mentioned Euler. There had to be another reason for his place on the list. Mr Ruche decided to go back to Euler's biography.

A few minutes later, Max came into the Rainforest Library, as he often had since Sidney's disappearance, and sat silently with Mr Ruche. Mr Ruche reopened Euler's *Complete Works* and decided he would read aloud from it.

'In 1760, during the Seven Years' War, Russian troops occupied part of Germany. As they passed through Charlottenburg, they ransacked Euler's house. When General Tottleben was informed, he sent a message to Euler which declared, "We have not come to make war on the sciences." '

''Course not, they didn't want to hurt any theorems, just kill some people. So what did Tottleben do?' asked Max.

'Euler was immediately reimbursed', said Mr Ruche.

'How do you reimburse a page of maths? How much is a theorem worth, Mr Ruche?'

Mr Ruche couldn't answer, so he carried on:

'The Empress of Russia, Catherine the Great, wanted Euler to work in her Academy of Science. Euler was happy to get away from Berlin, and from Frederick II, with whom he had a difficult relationship, and so he left for St Petersburg. Frederick II wrote to d'Alembert after the voyage, relating that "Mr Euler, who is a passionate astronomer, went to the north to look at the Great Bear and the Little Bear. The ship carrying his papers sank. Sadly, everything was lost, which is a great shame because there was enough to fill six volumes, and all of Europe will now be deprived of the pleasure of reading them.'

'It sank? What about Euler?' asked Max.

'He wasn't on the boat', said Mr Ruche. He walked over to the kettle and switched it on. He wondered how Euler had felt, having some of his manuscripts destroyed by Tottleben's soldiers, others lost at the bottom of the Baltic Sea. And this, Mr Ruche remembered, was before his library burned in St Petersburg. He made himself a cup of tea and went back to his reading:

'Euler rewrote from memory everything he had lost. He had a remarkable memory: one night he decided to calculate the first hundred numbers to the first six powers and learn them by heart. 51 to the power of 5, for example, or...'

Max didn't give him time to go on, he had already keyed it into his calculator and announced: '345,025,251.'

'Or, I don't know, 77 to the power of 6', said Mr Ruche.

'208,422,380,099', announced Max.

'He calculated all six hundred powers and learned them all by heart. Just thinking about it makes me feel dizzy. How could you possibly memorize all those numbers and still get to sleep? It's not even as though he was showing off – everything he committed to memory would help him in his work. He was the true successor of Fermat. He wrote over a hundred and fifty articles, and he knew every formula in trigonometry by heart. It wasn't just maths – he could recite the whole of Virgil's *Aeneid* from memory. He could even remember the first and last lines of every page in the edition he had read as a child!'

'It's his memory', exclaimed Max, 'that's what Grosrouvre was trying to get at, Grosrouvre's loyal friend could recite a whole text from memory – he could recite Grosrouvre's proofs!'

'Bravo Max! I think you've hit the nail on the head. It wasn't about the fire, it was Euler's memory!'

Max took the book from Mr Ruche and read aloud: 'At the age of twenty-eight, Euler was faced with a difficult astronomical problem. After three days of uninterrupted work he solved it, but the effort was such that he suffered a stroke and lost his sight in one eye. He knew that he would eventually be completely blind and decided to prepare himself.

'He began by learning to write as though he were blind. He would close his good eye, take a piece of chalk and write mathematical formulas on a large blackboard. As time went on, he could write long, complex analytic formulas without a mistake.

'Every day he tried to memorize mathematical texts so that when he could no longer read them they would still be there in his memory. He became a living library.'

A living library, thought Mr Ruche, which was exactly what Grosrouvre had made of his 'loyal friend'. Euler learned texts by heart so he would remember them when he was blind. Grosrouvre had persuaded his friend to learn his proofs by heart, not because his friend was going blind, but because the texts themselves would no longer exist. He was going to burn them.

Mr Ruche poured himself another cup of tea and sipped it slowly. 'Right', he said cheerfully, 'I think I'll just learn the Rainforest

Library by heart. That would be the best insurance against a fire!'

'Don't flatter yourself, Mr Ruche!' retorted Max, 'Euler developed a good memory because his eyesight was bad. It's usually because one sense is bad that you develop another to take its place. Anyway, it's just as well Euler learned all his books by heart. Even if he hadn't gone blind, his books were all burned in the fire anyway.' Then he added, 'That's what would have happened to the Rainforest Library if your friend hadn't sent it here.'

A terrible thought occurred to Mr Ruche. He had thought that Grosrouvre sending the books was a happy accident. Now, he realized, there was nothing miraculous about it. Grosrouvre had sent him the library because he knew that his house would be burned down. The fire wasn't an accident, it was arson. But Mr Ruche could not bring himself to believe that Grosrouvre had set his own house ablaze.

Max continued to read aloud: 'Euler's left eye soon began to fail. Shortly after his arrival in St Petersburg, he had gone completely blind. He decided to undergo surgery to have a cataract removed. The operation was successful, for his sight returned. It was the greatest joy in his life. He took great pleasure in writing his own letters again – to Bernoulli, Lagrange, Goldbach and many others...'

'What was that last name?' asked Mr Ruche.

'Goldbach.'

'Goldbach...Goldbach's conjecture – that was the other thing Grosrouvre said he had solved.'

'We can't do everything at once', said Max. 'Let's finish with Euler before we get started on Goldbach', and he continued reading.

'Euler's eye became infected and he lost his sight once more, though he continued to work, this time on a book about algebra. He employed a local tailor's son to take dictation. In order for the arrangement to work, the text had to be such that, in writing it, the boy would learn mathematics. By the time the book was completed, the tailor's son was capable of solving complex algebraic problems.'

'On 7 September 1783, he was drinking afternoon tea when he suffered a fit of apoplexy. What's apoplexy?'

'A heart attack.'

Max put the book down and stared at Mr Ruche: 'Mr Ruche, I don't want you to drink tea any more.'

CHAPTER 21

The Proof of the Pudding

Goldbach's conjecture; Fermat's second theorem; unrequited love & how Fermat's Last Theorem saved a man's life

Mr Ruche stared at Grosrouvre's letter: 'An unproved theorem – a conjecture – is a simple statement, something the whole world takes for granted, but which no one has been able to prove. That was the sort of challenge I needed!' He put the letter down on the desk, wheeled himself to Section 3 of the Rainforest Library and took out Grosrouvre's index card on Goldbach:

Goldbach's conjecture

In 1742 the mathematician Christian Goldbach sent a letter to his friend Leonhard Euler, in which he wrote 'every even integer greater than 2 can be represented as the sum of two primes'. For example, 16 = 13 + 3 or 30 = 23 + 7.

Gauss had already established that all whole numbers could be factorized into the product of an unspecified number of prime numbers. Goldbach was now affirming that they could all be decomposed into the sum of a specific number of prime numbers.

Two hundred and fifty years have passed, and we still have no proof that Goldbach's conjecture is true. I'm working on it.

There was another note, written in blue ink, which was clearly more recent:

N.b. The Russian mathematician Ivan Vinogradov has proved that all whole, odd numbers greater than $3^{14348907}$ are the sum of three prime numbers. Recently, the Chinese mathematician Chen Jing-run has made great progress on this subject, but the theorem remains to be proved. I think I am working along the right lines.

Mr Ruche brought the book back to his desk. As he read, he took notes:

> Increasingly interested in Fermat's work, Euler arranged to study his papers. In the middle of the proof that 'no right-angled triangle can have an area which is a square', he discovered in the margin of Diophantus' *Arithmetic* a proof of the theorem for $n = 4$:
>
> $x^4 + y^4 = z^4$ has no solution in whole numbers
>
> It was, in fact, the only time that Fermat had explicitly used the method of infinite descent. Using the same method, Euler immediately set to work to prove the theorem for $n = 3$, but he used complex numbers rather than real numbers. On 4 August 1753, he was able to announce that he had proved: 'The cube of a whole number cannot be the sum of two cubes.'

Grosrouvre had written on the index card:

> Euler's proof contained an error. His method, however, was flawless and was later used with great success.

That evening, Mr Ruche invited the whole family to an 'evening of conjectures'. Everyone realized the importance of the meeting, and they prepared to tackle the question they had been thinking about for months: had Grosrouvre really succeeded in proving Fermat's Last Theorem or Goldbach's conjecture? Mr Ruche gathered together all Grosrouvre's index cards on Fermat and Goldbach, and began to read:

> **Attempts to resolve Fermat's Last Theorem**
> First observation: it is sufficient to prove the theory true for prime numbers.
>
> Successive generations of mathematicians attempted to prove the theorem 'gradually'. Unable to solve it for the general case of all primes, they focused on specific cases which they could solve, hoping that in doing so they might stumble upon a solution for the general case. Legendre proved the theorem for $n = 5$, Gabriel Lamé proved it for $n = 7$ and Gustav Dirichlet proved it for $n = 17$.
>
> In 1820, a young woman, Sophie Germain, who, because of constraints on women writers at the time, had published some work

under the name 'Monsieur le Blanc', was the first to offer a general solution – but it was for a specific subset of prime numbers.

On 1 March 1847, a momentous session took place at the Academy of the Sciences. Two men, Lamé and Augustin Cauchy, stood up in turn and presented a sealed envelope each containing a complete proof of Fermat's Last Theorem. There was consternation in the hall – which of the two would take the gold medal?

A month passed. At the following session, a German mathematician, Ernst Kummer, demonstrated that both of them had mistakenly attributed the properties of real numbers to complex numbers. Both proofs were therefore false. They had made precisely the error Euler had a century earlier.

Shortly afterwards, Kummer, relying on the properties of what he called *ideals*, proved the theorem true for most of the primes less than 100.

In the second half of the twentieth century, things speeded up a little. With the help of computers it was possible to prove the theorem true for thousands and, in time, hundreds of thousands of primes. But each proof dealt only with a finite number of primes. The theorem would not be proved until it had been proved true for *all* primes!

Jonathan said: 'I'd like everyone to note that one of the greatest mathematicians of the nineteenth century *thought* he had proved Fermat's Last Theorem, and he was wrong.'

Mr Ruche calmly picked up the next index card and began reading:

At the beginning of the nineteenth century, all the problems left by Fermat had been resolved except one. The sole remaining problem was his 1637 conjecture, which became known as Fermat's Last Theorem. There was a certain irony in the title, in that it was not a theorem. That was precisely the problem, in fact – it would not become a theorem until it had been proved – if it ever was. The more difficult it proved to solve, the more famous it became.

Just before the First World War, a prize was established by Paul Wolfskehl, a rich German, for the first person to prove it. The prize money was considerable, but there was a condition: Fermat's Last Theorem had to be proved before 13 September 2007.

One more thing needs to be taken into consideration. In 1640, Fermat proposed his second theorem in a letter to Frénicle: 'I am persuaded that (2 to the power of 2^n)+1 is always a prime number. I have not succeeded in proving it completely, but I have managed to exclude so many possible factors and my thoughts are so clear that I would find it difficult to believe it untrue.' In 1732, however, Leonhard Euler showed that Fermat's fifth number, (2 to the power of 2^5) + 1 (i.e. $2^{32} + 1$), equals 4,294,967,297, which is divisible by 641 and therefore not prime. Fermat's second theorem was false. If he was wrong once, surely he could be wrong again.

Jonathan said: 'I'd like everyone to note that one of the greatest mathematicians of the eighteenth century *thought* he had proved Fermat's second theorem, and he was wrong.'

Mr Ruche continued:

This is why, aware of the countless attempts made by mathematicians before me, each of them convinced of the veracity of the theorem, I began by attempting to prove that it was false. I spent a long time in the attempt, but my work simply served to convince me that it was indeed true, having succeeded in proving that, in a number of respects, it could not be false. Since then, I have been working to prove this.

'There's just one thing', Perrette asked, 'why did the rich German choose that date?'

'13/9/2007? 13 is a prime number', said Jonathan, 'though 9 isn't, and 2007...I don't know – maybe it's a prime number too.'

'No it's not', interrupted Perrette. 'When I was little, I was taught that if you could divide the sum of the digits of a number by 3, then the number could be divided by 3. 2 + 0 + 0 + 7 equals 9, which can be divided by 3.'

'I'm afraid it's simpler than that', said Mr Ruche. 'Paul Wolfskehl was rich but unhappy, because he was in love with a young woman who didn't love him.'

'You wouldn't catch me loving some woman who didn't love me', said Jonathan.

'It's not quite that simple', said Lea, laughing.

'Could you love someone who didn't love you?'

'The question doesn't arise – everyone loves me', said Lea.

'Paul Wolfskehl was so unhappy that he decided to commit

suicide. He chose a date. When his last day came, he organized his papers and wrote a will. When he had finished, he realized that there were still two hours to go before midnight. He went to his library, thinking that he would read something to pass the time. He took down Ernst Kummer's book on Fermat's Last Theorem. As he was reading, he thought he found a mistake. He looked at the clock – he still had time. If he could spend his last hours on earth proving that there was a mistake in the work of such a great mathematician, what a wonderful end it would be.

'He sat down at his desk and began to work through Kummer's text, line by line. When he came to the end, he realized that it was correct after all. There was not the slightest mistake. Disappointed and exhausted, Paul looked up from his paperwork. Midnight had long passed. Dawn was breaking and he was still alive!

'He closed Kummer's book, folded his papers, put away his pistol, tore up his will and forgot about the woman he loved. He had found a solution to his unhappiness – mathematics! He felt he owed a debt to Fermat and to his Last Theorem, and decided to create a prize to reward whoever should solve the problem that had saved his life. Paul Wolfskehl had decided to kill himself at midnight on 13 September 1907!'

Mr Ruche picked up the one card left. Grosrouvre had clearly written it recently.

Euler's conjecture

Extrapolating Fermat's conjecture, that the sum of two numbers to the *n*th power cannot be a number to the *n*th power: $x^n + y^n \neq z^n$, Euler proposed a more modest conjecture using four numbers rather than three, and limiting it to the 4th power: 'The sum of three numbers to the 4th power cannot be a number to the 4th power.'

Nowadays, it would be expressed as '$x^4 + y^4 + z^4 = w^4$ has no solution in whole numbers'.

The conjecture lasted two hundred years. In 1988, the mathematician Noam Elkies found four numbers that contradict it:

$$2{,}682{,}440^4 + 65{,}639^4 + 18{,}796{,}760^4 = 20{,}615{,}673^4.$$

Euler's conjecture was false!

Mr Ruche was shocked. Euler, the celebrated mathematician, who had written 75 of the volumes in the Rainforest Library, had proposed a conjecture that was false. Why was Grosrouvre so insistent in pointing out the errors of great mathematicians, wondered Mr Ruche.

'I'd like to point out', began Jonathan, 'that one of the greatest mathematicians of the nineteenth century...'

'We heard you the first time!' shouted everyone.

CHAPTER 22

In Mathematics, Nothing is Impossible

The three great problems of Greek mathematics & a nasty surprise

Jon-and-Lea, who were revising for their exams, looked up as Mr Ruche's voice boomed from the lift in the courtyard:

'The Royal Academy for the Sciences, Paris, 1775. The Academy has decided that it will no longer examine solutions to the problems of duplicating a cube, trisecting an angle or squaring a circle, nor those of any machine said to have perpetual motion.'

He wheeled himself inside, waving a photocopy. 'I found this at the National Library. Seventy years of experience had proved to the Academy that – and I quote – "none of those who sent solutions understood the nature or the difficulties of these problems, and none of the methods they employed were likely to lead to a solution, if one were possible."

'So the Academy concluded that it was pointless to study all the "solutions" sent to them. At the time, there was a rumour that the government had promised a hefty reward to anyone who succeeded in squaring the circle. Hordes of people gave up their jobs to search for a solution to the problem.

'Some of them really believed they had solved it and refused to accept the flaws in their proofs. Therefore the Academy, convinced that further investigation into such claims was pointless, made a public declaration, hoping to quash the rumours which had proved the ruin of more than one family.'

This last phrase echoed in the silence that followed, and Perrette wondered if Mr Ruche was implying that attempting to solve Grosrouvre's problem might ruin them. The only thing that had happened since they began their research was the kidnapping of Sidney – but that had nothing to do with Grosrouvre. 'Can you read the sentence that begins "Seventy years of experience..." again?' she asked.

Mr Ruche reread the passage. When he came to the phrase 'and none of the methods they employed were likely to lead to a solution, if one were possible', Perrette interrupted:

'That's it! That's what I thought! The Academy was saying that there might not be *any* solution to the problems.'

'What?' said Jon-and-Lea together. 'You mean the three problems...'

'Hang on, hang on! Not so fast!' said Mr Ruche.

'That would mean that all the mathematicians of the ancient world...'

'And the modern world', added Jonathan.

'...had wasted their time trying to solve problems that have no solution.'

'You're jumping to conclusions', said Mr Ruche. 'It doesn't say there is no solution, it says "if one were possible".'

'I'm sorry, Mr Ruche', said Lea, 'but that's just it: it says a solution might be possible, not that it might be impossible. The mathematicians in the Academy obviously didn't think the problems could be solved.'

Mr Ruche nodded. Lea was right: the Academy clearly felt that these problems had no solution. When had mathematicians stopped trying to find solutions for everything, and begun accepting that some problems might not have them? For thousands of years Greek and Arab mathematicians and all those who followed them believed these problems could be solved. What had changed their minds?

Mr Ruche spent the afternoon with Max researching the history of the three great problems of Greek mathematics. He had first mentioned them just before Christmas. Now, he decided he would present them again. That evening, everyone gathered in the Rainforest Library. Knowing that the session was important, Perrette had closed the bookshop early. In the library the curtain had been drawn, but this time there was no stage set.

'Let's start with squaring the circle', Mr Ruche suggested. 'In the middle of the sixteenth century, a German mathematician,

Michael Stifel, was the first to suggest that squaring the circle might be impossible. His statement had little impact. The number of mathematicians attempting to square the circle grew every year. Each new attempt produced new errors, but amateur mathematicians, rather than being discouraged, simply saw the opportunity to try again. It was like a competition.'

Mr Ruche gestured to Max, who announced: 'A journey through the universe of numbers!'

'Thanks to Tartaglia, Cardano, Ferrari, Bombelli, Abel and Galois, we have spent a lot of time studying algebraic equations. This will help us to define a new property of real numbers.'

The motor on the overhead projector purred, and a definition appeared on the wall:

An algebraic number is a number which is the solution to a mathematical equation.

Mr Ruche continued: 'Whole numbers, both positive and negative, are algebraic. -1, for example, is the solution to the equation $x + 1 = 0$. Rational numbers are also algebraic. For example, $\frac{2}{3}$ is the solution to the equation $3x - 2 = 0$.'

Max went on, 'But that's not all! $\sqrt{2}$ is also algebraic!'

'It's the solution to the equation $x^2 - 2 = 0$, which poses the question...'

A question appeared on the wall:

Are all real numbers algebraic?

'In other words', explained Mr Ruche, 'are there any numbers which are not algebraic?'

'What has this got to do with squaring the circle?' asked Lea.

'Patience!' said Mr Ruche. 'Since, as we've seen in the case of $\sqrt{2}$, some irrational numbers are algebraic, it is natural to wonder whether all irrationals are algebraic.'

Another slide appeared:

Are there any irrational numbers that are not algebraic?

'Without knowing whether such numbers existed, mathematicians called them transcendental. Remember the adjectives that mathematicians used to describe numbers before: broken, fractured, imaginary, complex, ideal and now transcendental. Simply imagining the existence of transcendental numbers allows real numbers to be split in two separate ways.'

The next slide showed how:

Rational/Irrational
Algebraic/Transcendental

'Aside from common numbers and their roots, what other numbers did mathematicians use? There were π, e and logarithms, sine and cosine. Was π rational or irrational, algebraic or transcendental? Mathematicians in the eighteenth and nineteenth centuries tried to establish the links between these two categories of number. This was one of the questions that preoccupied them.

'Now, Lea, you asked what all this has got to do with squaring the circle. Let me start by pointing out one of the important differences between the square and the circle. It is easy to show that the ratio of the perimeter of a square to its diagonal, $2\sqrt{2}$, is irrational, but it's difficult to prove that the ratio of the circumference of a circle to its radius, π, is irrational.'

Another slide made the point:

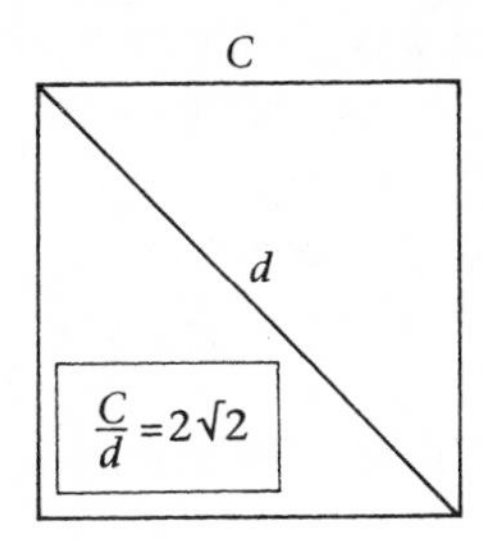

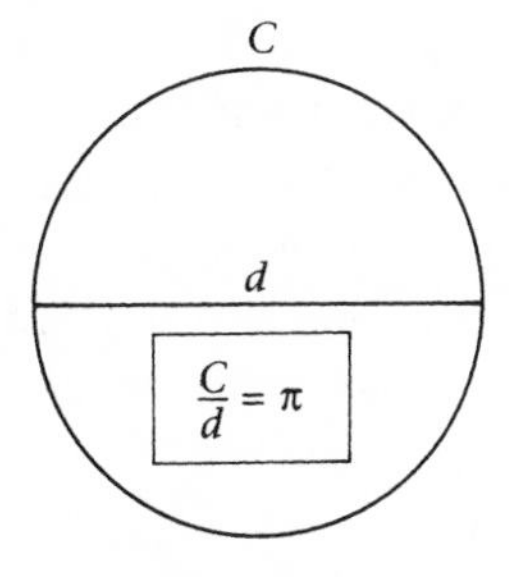

'This is where Leonhard Euler comes in. He was the first to conjecture that π was not only irrational, but also transcendental. He was unable to prove it. Some years later, in 1761, Heinrich Lambert came up with a proof.'

'Johann Heinrich Lambert was rather an odd fellow. One day, he asked to be received by King Frederick II of Prussia, who asked him:

'"What do you know?"

'"Everything, Your Majesty!" Lambert replied.

'"And from whom did you get this information?" asked the King.

'"From myself, Your Majesty." said Lambert.'

Another slide appeared on the wall:

π is irrational

'So π isn't equal to $\frac{22}{7}$ after all?' asked Perrette naively.

'Absolutely not!' cried Mr Ruche with a look of horror on his face. 'If π were equal to $\frac{22}{7}$, it wouldn't need a name, we'd call it $\frac{22}{7}$ like any other fraction. Squaring the circle would be easy – and the Temple of π in the Palace of Discovery would never have been built.'

'And billions and billions of decimal places would be out of a job', said Jonathan. 'See what you've done, Mum?'

Mr Ruche explained that Lambert, who 'knew everything', had not succeeded in proving π was irrational, and neither had his successor, Adrien Legendre, who had, however, succeeded in proving that π^2 was irrational.

'Two crucial changes occurred in the way the problem of squaring the circle was viewed. The first was moving from the certainty that the circle could be squared, to considering that it might prove impossible. The second was the shift from geometry to algebra. After two thousand years of attempting to square the circle by geometric methods, if was time to "algebrize" the problem.

'This was accomplished by a young student at the École Polytechnique. Laurent Wantzel was only twenty-three in 1837 when he developed a theorem that demonstrated the types of equation which could not be solved with a ruler and compass. It was to have profound consequences.'

Mr Ruche paused for effect before declaring, 'The duplication of the cube was just such an equation.'

A slide appeared to reinforce the point:

The duplication of the cube is impossible with a ruler and compass.

Jonathan, Lea and Perrette were completely engrossed. The next slide appeared as Mr Ruche made another declaration:

'The equation for the trisection of an angle was just such an equation.'

> *The trisection of an angle is impossible with a ruler and compass.*

Max spoke up: 'So two of the three problems of antiquity had been proved impossible.'

'What about squaring the circle?' asked Jonathan impatiently.

Mr Ruche ignored the question and continued: 'In 1882, the German mathematician Ferdinand Lindemann finally proved that π was transcendental, which meant that π could not be the solution to an algebraic equation. The search for ways to square the circle was over!'

A new slide appeared:

> *Squaring the circle is impossible with a ruler and compass.*

The three declarations together seemed impressive. Perrette had been right when she thought the session would be an important one.

Max concluded, 'It took two thousand four hundred years to prove that the three problems of Greek mathematics are impossible to solve with a ruler and compass.'

The room was completely silent. Everyone was thinking about what they had learned. Since the beginning, they had identified these three problems with the questions surrounding Grosrouvre's death. Would these prove impossible to solve?

Perrette closed the shop at five o'clock the following afternoon. When Max was still not home from school at seven, she started to worry. She phoned the school, but there was no answer, so she raced over to catch the caretaker before he left. He phoned the headmaster, who told her that Max had left with his friends at 4.30 p.m., as usual. She stopped at the corner shop as she walked home to ask Mr Habibi whether he had seen Max.

'Yes, I saw him', said Mr Habibi. 'He waved to me, and then he was gone.'

Perhaps Max was home by now. Perrette ran all the way back. Mr Ruche was waiting for her outside the bookshop. When she saw his face, she was terrified.

'They've kidnapped Max', he said, his voice trembling.

'How do you know?'

'They phoned.'

'Who? Who phoned?'

'I don't know.'

'We have to call the police.'

'No, Perrette! They said they'd hurt him if we called the police. They'll call back tonight.'

'We should have phoned the police when they took the parrot.'

The phone rang, and she ran into the bookshop and grabbed it.

'Hello? Hello? What have you done with...Oh, Jonathan... Jonathan, they've taken Max.' Perrette started to cry. Mr Ruche took the handset from her and told the twins what had happened.

'It's OK, Perrette. They're on their way.'

The telephone rang again. Perrette grabbed it before Mr Ruche could stop her. He watched the colour drain from her face.

'Hello? Who is this? Who is this?' She handed the phone to Mr Ruche. 'They want to talk to you.'

'Hello', he said calmly. 'No, I assure you, no one has called the police.'

'Good, now listen up!' snarled the voice at the other end.

Mr Ruche listened in silence, and Perrette leaned close so that she could hear. When he hung up, she looked at him, stunned at what he had agreed to.

'You're not seriously thinking of going?'

'Of course I am!'

'You're mad! How can you go to Sicily? Max is my son, I should go.'

'Perrette, I have to...I don't think you understand.'

'And you do? Someone steals a parrot from my house, my son has been kidnapped in broad daylight, and now they're saying you have to go to Sicily. I don't understand! Explain it to me!'

'Perhaps I don't understand any better than you do, but I do know one thing – these men are serious. I don't think they'll hurt Max as long as I do what they say. They said Max was on his way to Sicily already.'

'The Mafia – it has to be them! But what do the Mafia want with Max, or the parrot, or you?' She looked at Mr Ruche and began to sob again, her whole body shaking.

Mr Ruche tried to smile. 'It's OK, Perrette, don't cry. I don't understand what's going on either, but sometimes you just have to do what you can. Now, I have to get some things together. I have to leave tomorrow.'

Perrette helped him pack.

Neither of them could sleep, so they sat up all night talking with the twins. Late that night, on his way back from the airport, Albert noticed all the lights on and called in, worried that something had happened to Mr Ruche. When he heard about the kidnapping, he sat in silence as they talked about what to do.

The following morning, Mr Ruche phoned the travel agent to book a plane ticket and discovered there was a general strike in Italy. All public transport was affected, including air traffic control. The strike looked set to last for several days.

He was still absorbing this information when Albert fumbled with his cap, lit a cigarette and said, 'I'll take you.'

'But you can't drive all the way to Sicily!'

'You don't think the Renault is up to it?'

'I don't know, but...what about your work?'

'I've always liked being me own boss. You can leave whenever you want. Anyway, I'd like to go back to Italy, I didn't see much of it last time I was there, and I've always wanted to see Syracuse.'

As the Renault pulled away, Perrette and the twins stood waving, hoping against hope that Mr Ruche and Albert would be back soon. Jon-and-Lea knew that their long-awaited trip to Manaus was looking more and more unlikely. If Max came back safely...*when* Max came back safely, they could think about it. They were still convinced that going there was the only way to solve the mystery of Grosrouvre's death and the missing theorems.

As they walked back upstairs, they heard the telephone ringing in the bookshop. They rushed downstairs and Jonathan grabbed the phone. A voice on the other end said 'Hello, Mum!' It was Max! Jonathan smiled and handed his mother the phone.

Max explained that he had found Sidney and that they were both fine. He told her not to worry, that he loved her, and to give his love to Jon-and-Lea and to Mr Ruche.

Perrette waited for him to finish, then told him that Mr Ruche had already set off with Albert and they would be in Sicily in a few days. As she said it, she realized that of course he couldn't hear her. Suddenly, she didn't know what to do. Then a woman's voice said, 'I've explained to Max what you said. He seems to be happy. You have a very nice son, madame.'

Then she hung up.

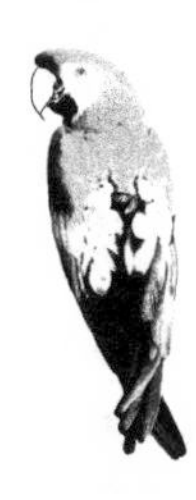

CHAPTER 23

The Third Man

The third member of the trio; a secret gallery & Grosrouvre's loyal friend

Albert parked the Renault in front of a bar on the Porto Piccolo in Syracuse and helped Mr Ruche inside. Before they had said a word, the barman handed Mr Ruche a note giving him directions to the Orecchia di Dionisio – the 'Ear of Dionysius'. As soon as they left, the barman picked up the telephone.

They drove through the city towards the Neopolis Archaeological Park. As they drove, Albert talked nervously, telling Mr Ruche what he had learned from his passengers over the years. The Greek amphitheatre, he explained, was the largest in the ancient world. It had been dug into the hill itself and could hold an audience of 15,000. When the Romans invaded they took it over. In other circumstances, Mr Ruche would have wanted to know more, but he was worried and frightened.

The Ear of Dionysius turned out to be a cave in the huge quarries which surrounded Syracuse and had provided the stone from which the ancient city had been built. They parked the Renault on a verge overgrown with wild pomegranates, lemons and oranges. Albert got out and walked around the car, but there was no one in sight. He got back into the car, where Mr Ruche was sitting in silence.

'They say that Dionysius used to imprison people in the caves here', said Albert. 'The acoustics amplified a man's breath so that it sounded like a storm. Dionysius used to press his ear to the cliff-face to hear what the prisoners...'

'Get out of the car and tell the driver to leave!' a voice thundered from inside the cave. Albert dropped his cigarette.

'I warn you, there is a gun trained on you.'

'Do as he says, Albert', said Mr Ruche quietly.

'Go back to the bar on the Porto Piccolo', the voice continued. 'You will be given the address of a hotel where you will stay until contacted. If you say anything to anyone...'

Albert helped Mr Ruche out of the taxi and put his suitcase and one that Perrette had packed for Max beside him. Then he got back into the car and drove off.

Mr Ruche was staring towards the Ear of Dionysius when a sudden sound made him turn. A van pulled up close to him, and a man got out. Had Albert glanced back at that moment, he would have recognized the man as the Tall Stocky Guy, who he had refused to take to Roissy Airport. The back door of the van opened and a ramp was lowered automatically. Mr Ruche's wheelchair was pushed into the van.

The Tall Stocky Guy drove in silence. After a quarter of an hour or so, the van stopped outside a large mansion. He spoke gruffly into the entry-phone and the gates opened, closing again silently after they had driven through. Two guard dogs, growling and yapping, ran alongside the van as it moved up the long driveway.

From the balcony outside the mansion, an old man watched the van as it approached. The dogs arrived first, bounding up to the balcony. At a wave of his hand they stopped and lay down on the gravel path. The sun was still high in the sky.

The Tall Stocky Guy rolled the wheelchair from the van and pushed it to a shady spot under an orange tree. Mr Ruche watched as the old man walked towards him. He was tall and handsome with a shock of white hair framing his hard, thin face. He was elegantly dressed in a white linen suit and carried a handsomely carved walking stick. Despite his age, he had a grace about him that made him all the more imposing. He stopped some feet away, put on his glasses and studied Mr Ruche.

'My God!' he said softly.

'I want to see the boy!' Mr Ruche demanded. 'Now! And if you've touched a hair on his head...' His face was hard with rage. The man gestured to the Tall Stocky Guy, who nodded curtly and walked back to the house.

'You don't recognize me?' asked the old man.

'I'm afraid I don't know you, and I don't wish to.'

'Well, I recognize you, Pierre, even after all these years.'

Surprised, Mr Ruche studied this man who appeared to know him. The man gestured with his walking stick.

'Pierre Ruche, the philosopher. I'd know your face anywhere.'

'Tavio?' Mr Ruche's voice was tremulous with shock. He stared in disbelief at Tavio, the waiter from the café across from the Sorbonne. He was confused to find himself face to face with someone he had not seen for half a century.

'Tavio', he half-whispered, 'it can't be...what are you doing here? Did you have me brought here? What the hell is going on?'

Suddenly, it seemed that Perrette had been right. Tavio was the third man: Grosrouvre, Ruche and Tavio had been a trio. Now, in some way he could not begin to understand, Tavio was involved in this mystery. Mr Ruche sat up in his wheelchair:

'You were the one who had Max kidnapped! Have you gone mad? He's a child, for God's sake – he's only twelve!'

There was the sound of footsteps on the gravel. Max ran to Mr Ruche as fast as he could and threw his arms around him.

'It's OK', Mr Ruche hugged Max to him, 'it's OK. Are you all right? They didn't hurt you, did they?' He was crying. He hadn't cried for more than twenty years.

Max felt a tear fall on his hand, and whispered into Mr Ruche's ear, 'They're watching us.'

Mr Ruche let Max go.

'They didn't hurt you?' he asked again.

'No, I'm fine', said Max, 'and Sidney's fine too.'

'You see', said Don Ottavio, 'I'm a civilized man.'

Mr Ruche said nothing. He was thinking about what Perrette had said when he had told the story of the three friends. She was right, Grosrouvre had been trying to remind him of the third person in the trio.

Mr Ruche looked hard at Tavio and everything clicked into place. This was what Grosrouvre's clues were pointing to. Tavio was the one who wanted Grosrouvre's proofs, thought Mr Ruche, he thinks Grosrouvre sent them to me. He had Max kidnapped to force me to give them to him. But what did he want with Sidney? The long journey had worn him out. It was hot, even in the shade of the orange tree. Max seemed to be unharmed, that was the important thing. In any case, Mr Ruche didn't know anything about Grosrouvre's proofs and he was tired and confused. He felt his body go slack, and was aware of Tavio running towards him and Max shouting. Tavio threw aside his cane and caught Mr Ruche just as he fainted and fell from his wheelchair.

When he opened his eyes again, Mr Ruche did not know where he was. He was lying in a bed that felt impossibly soft. The bed was in the shape of a boat, the prow pointed towards the window, through which Mr Ruche could see a blue strip of sea. The room was vast, with a magnificent dresser that served as a bookcase; panes of leaded glass allowed glimpses of the books within. Mr Ruche felt better now. It was growing dark. Outside, he could hear the voice of Don Ottavio.

Mr Ruche could hardly believe that Tavio had become a feared gang leader, but this was Sicily – perhaps he was a Mafia boss. Don Ottavio turned towards the window and Mr Ruche closed his eyes quickly. He needed time to think.

Although Mr Ruche still did not understand why Sidney was involved, he was convinced that it had something to do with Grosrouvre and his proofs. He decided that the best course of action was to explain everything to Don Ottavio. He would tell him about the letters, the library, everything. He would tell Tavio that Grosrouvre had not sent him the proofs. Mr Ruche hesitated – perhaps Grosrouvre had hidden the proofs in one of the books in the Rainforest Library? It was strange that he hadn't thought of this before. But if he had, how could Mr Ruche betray him? He deserved better than that. The more he considered the problem, the more complicated it seemed. It was like trying to unravel a tangled ball of string – the more you try to untangle it, the more tangled it becomes. But at least if he told Tavio everything, he, Max and Sidney could leave for the hotel and find Albert, who, Mr Ruche knew, would be sick with worry. He was about to call out to Don Ottavio when he remembered something he had learned in the Resistance: a prisoner holds all the cards, all the information. He had been taught to say nothing and never try to anticipate the jailer. He shut his mouth and reconsidered his decision. He would say nothing.

Don Ottavio knocked and came into the room, followed by a

man in a dark suit who introduced himself as a doctor. At first Mr Ruche refused to be examined, but Max insisted and, eventually, he agreed. The doctor took his pulse and his blood pressure.

'Your friend is in excellent health', he concluded. As he was leaving, he added, 'He has the heart of a man half his age.' The doctor blushed and looked apologetically at Don Ottavio.

'My heart is weak', explained Tavio. 'Sometimes it yaps like a small dog. Don't worry about the boy, he's fallen asleep. Now you must rest. We shall talk later.'

Mr Ruche woke at dawn and watched the sunrise through the open window. At eight o'clock a maid came to help him get ready. Max was still fast asleep on the bed at the far end of the room. 'Don Ottavio would like you to have breakfast with him', said the maid. She wheeled him into a small dining room where Don Ottavio was reading a newspaper. When he heard them coming he took off his glasses and said good morning.

'I'm glad to see you're feeling better. We were very worried about you. Come, have some breakfast.'

Mr Ruche could feel his resistance falter. He had to speak now.

'Tell me why you had Max kidnapped!' he demanded tersely. 'And what did you want with the parrot? What exactly do you want from us?'

'I shall answer all your questions', Don Ottavio replied calmly, 'but first I should tell you that Elgar is dead. He died almost a year ago in a fire in his house in Brazil.'

Mr Ruche didn't blink. Then, as though he were struggling to remember something, he said, 'I thought he died years ago. What on earth was he doing in Brazil? What has this got to do with my question?'

'I shall try to explain', said Don Ottavio. 'If you remember, we met just before the war. I was seventeen. My family had moved to France some years before. We come from a village in the mountains above Etna.' He pointed to the hills. 'My father was a mason but he could not find work here so he decided to emigrate. His brothers had already moved to New York, and they invited him to join them. They said they could find work for him.'

Don Ottavio gestured to a butler who offered Mr Ruche some orange juice.

'The oranges are grown on the estate', said Don Ottavio, pouring himself a coffee. He sipped it slowly before continuing. 'My father refused. He always hated the sea. He was sure that the journey to America would kill him. I wanted to stay here, but I could not argue with my father. I went with the family. I was about eleven, maybe twelve, like your boy?' Mr Ruche nodded. 'So we left for France.'

'My father worked in the mines in the north of France. I did odd jobs here and there before I moved to Paris and got some work in a café. That was where I met you and Elgar. I was so jealous of you: "Being and Nothingness" remember? And we became friends. You used to take me to parties in the Latin Quarter. It was through you that I met my first girlfriend. In the afternoons, Elgar would come and sit in the café and work. I would often sit with him, and he would talk to me about maths. I didn't really understand much, but I listened. He was brilliant.

'Then the war came. You were both called up. Elgar wrote to me once. He told me he had been injured – his leg – and that he didn't know where you were. I thought you had been killed.

'My father was ill. His lungs had been damaged from working in the mine. When it got bad, he wanted to come back to Italy. He didn't live to make it back to the island. I suppose at least he was spared the boat trip.' Don Ottavio laughed bitterly.

'But I came home, with my mother and my brothers. When I got here I joined the Resistance. Then the Americans landed here and peace came. My uncles in New York began to send me contraband cigarettes. I made money – a lot of money! That was the first step to my becoming Don Ottavio. I became rich enough to buy this mansion. I could buy anything, and I did. Houses, horses, cars...and women. Everything has a price.'

It all sounded so different from Mr Ruche's life. Don Ottavio then told him how he had run into Grosrouvre again. He had been on 'business' in Manaus to meet a client. It was there, in a café, that he had met his old friend again.

'He had become quite a rich man. We worked a little together. Some books he wanted. I suppose you would call it trafficking.' Suddenly Tavio asked, 'Does the name Goldbach mean anything to you?'

Mr Ruche was surprised. He hesitated for a moment. 'It's German, isn't it?'

'Yes', said Don Ottavio, 'but do you know what it means?'

'Goldbach? *Gold...bach* – it means "golden stream".'

'The golden stream – in the Amazon there are many rivers full of gold. Elgar knew all of them. He was one of the biggest traffickers in the business.'

Don Ottavio had been back to Manaus frequently, he said, sometimes on 'business', but more often to see Grosrouvre. 'He had started working on his mathematics again. He used to say he needed to do it – maths was like a drug to him. It was good for him!'

'Good for him?' queried Mr Ruche.

'Well, he was eighty-four when he died.'

'We were the same age', said Mr Ruche coldly.

'I invited him to come and stay here. He could have taken things easy, brought his books and everything. The weather there was not good for his health. But he refused. He worked like a madman – he would sit down to work after dinner and not look up before dawn.

'He started to lose wight. I thought he was ill, and told him so, but he wouldn't talk about it. He was obsessed with his work, and after a while I became curious about it. For a long time he kept his work secret, but one night, when we had been drinking, he told me he had solved two theorems that some of the greatest mathematicians had spent centuries trying to crack. He said the second one was by someone called Goldbach – I laughed and asked if he had chosen it because of the name. He looked at me, surprised. He hadn't even thought about it until I mentioned it. Stream of gold! He decided to keep his proofs a secret. He didn't have to explain why – I understood. Do you know why?'

Don Ottavio nodded to the butler to leave them, then walked to an oval mirror on the far wall. He put one hand on either side as if to straighten it, and the wall began to move. A secret door opened, though from where he sat Mr Ruche could not see into the room beyond. Don Ottavio motioned to Mr Ruche to follow him. The doorway was narrow, but the wheelchair passed through easily. Don Ottavio pressed another mirror, identical to the one in the living room, and the door closed. The room was dark, the only light coming from a skylight in the centre of the ceiling. Don

Ottavio pressed a button, and concealed lights lit up the room. Mr Ruche gasped. He turned his wheelchair around and stared at a dozen paintings hanging on the bare walls – all of them old masters.

'Each and every one of them is stolen', said Don Ottavio, leaning on his cane, smiling. 'They are wanted by every police force in the world. There are huge rewards for anyone who can find them. Almost as much as I spent acquiring them!' He pointed to each in turn: '*View of Delft* by Jongking. *The Love Letter* by Vermeer. *The Flight into Egypt* by Rembrandt. Goya's *Duke of Wellington*. That diptych is from the school of Giotto. Rodin's *Portrait for My Father*. *L'Estaque* by Braque, and two Picassos: *Guitar and Fruit Dish* and *Child with a Doll*. And there, my favourite: *The Flautist* by Vermeer. It has just arrived from Tokyo. I had it authenticated only yesterday.'

He put on his glasses to study it. Mr Ruche stared at the paintings.

'Since you're so rich, why didn't you just buy them?' he asked.

Don Ottavio answered with a laugh. He went up close to *The Love Letter* and looked at it tenderly. 'Buy them? Like a car or a washing machine?' He frowned contemptuously. 'Exactly what I would expect a shopkeeper to say. Firstly, because most of the paintings were not for sale. They are national treasures, as they say. But that is not the reason.' He took off his glasses and put them away. 'You don't wear glasses?'

'No', said Mr Ruche, a little proudly, 'never.'

'Why did I not buy them?' He was mocking Mr Ruche now. 'To have something unique, that no one in the world has and that everyone wants, is satisfying, but it's easy. It's bourgeois. It's like a child flaunting a toy in the playground to make other children jealous. No, I wanted something else, something more sophisticated. I wanted to have things that are unique and to be the only one who knew I had them. I felt it the first time I bought a painting – it was stolen from the Rijksmuseum.

'Have you ever wondered why famous paintings are stolen? What can the thieves do with them? Who can they sell them to? They sell them to collectors, who hang them in secret rooms and admire them in private. A rich man buys a painting at an auction and hangs it in his living room so that his guests can admire it and he can listen to their mewling praise. When the visitor looks admiringly in his direction, he looks down humbly and the painting is forgotten. When you take possession of a painting, intimately, as I do, it is like making love in secret to the most beautiful woman in the town. When you next meet her on a crowded street, you act as though you barely know her.'

Mr Ruche was disoriented by this revelation, and took a moment to collect his thoughts. 'But you still haven't answered my question. Why have you brought us here?'

'I shall tell you. Once I discovered that Grosrouvre wished to keep his proofs secret, I needed to have them. I wanted them for the same reason I wanted these paintings.'

Mr Ruche was angry. 'You think you can own a mathematical proof in the same way as you own a Rembrandt?' His voice was condescending. 'How do you know the paintings you have are the originals and not some clever forgeries by an old copyist?'

His voice cold, Don Ottavio whispered, 'If someone were to do such a thing he would not live to boast about it.'

'That's not the point. The Vermeer painting you just received – *The Flautist*. You had to have it examined to prove it was authentic. You might know something about art, but you couldn't authenticate it yourself, so you took it to an "expert" who told you it was not a forgery. You have only his word for its authenticity.'

Don Ottavio seemed genuinely intrigued by Mr Ruche's point. 'That's true, but I still don't see what you're getting at.'

'It's simple. If you were to have Elgar's proofs, how would you know they were correct? They could be the ravings of an old man.'

'How can you call Elgar's work "the ravings of an old man"?'

'I didn't mean it literally. But hundreds of mathematicians – including the most famous in the world – have tried to prove these theorems and failed. I'm sure many of them thought they had succeeded. Why wouldn't Elgar be the same? Only a mathematician – and a great one – would be able to confirm they were right. But as soon as you had let him study them, he would possess them just as much as you would. More so, in fact, because he would have understood them. He could publish them if he wanted.'

Don Ottavio was livid: 'There's a little expression in Sicily: "Tombs don't talk." '

Mr Ruche stared at him. 'What do you mean?'

'A little joke', said Don Ottavio, 'I just wanted to point out that there is a solution to every problem.'

Mr Ruche felt shaken. This was no school debating society. They were not simply scoring points, they were playing with human lives. He had to seize the advantage. He had to convince Don Ottavio that his search for the proofs was doomed to failure.

'Everything you have owned until now has...has had a physical presence – cars, property, horses, paintings – even the women you have known have a physical body.'

'What a strange comment!'

'But in maths, there is no physical presence, simply ideas. You could never own those proofs, Tavio. You might as well give up.'

'You've forgotten something, my philosopher friend: I am convinced that Elgar's proofs are correct. That is proof enough for me. You should be relieved – I will not have to kill a great mathematician to confirm it.' Don Ottavio's voice became more urgent. 'But all this talk – I still don't have the proofs.'

He went to the mirror and twisted it. The secret door opened and Mr Ruche wheeled himself out. Don Ottavio turned off the lights and followed him, the door closing behind them like the lid of a sarcophagus.

The table had been cleared and the curtains were drawn. Don Ottavio suggested they take a walk around the grounds before it got too hot. Mr Ruche was still shocked by what he had discovered.

'Aren't you afraid that I will go to the police?'

'No. By the time the police had arrived and managed to find the room, the paintings would be miles away. In any case, I have every faith in the local police.' He smiled. 'The commissioner is a good friend of mine.'

They went out into the vast gardens. Mr Ruche wheeled himself into the shade. It was cool under the trees. He looked up. The foliage was so thick that no sunlight came through.

Don Ottavio watched him, and suddenly said, 'I thought to myself, Elgar must have left some record of his proofs. I can't believe that he worked for ten years to see all his ideas turn to dust. So I began to ask myself what sort of record he would have left. I was sure he would not have left a written copy, but perhaps a computer disk, or a tape recording, a video or a microfilm. I wondered how could he have left a copy of his work but kept it hidden. But, as you explained just a moment ago, mathematical proofs are not physical objects, and Elgar could not risk leaving a physical object since there was always the risk that it might be found.' He paused. 'Look at him – he certainly has an appetite!'

Mr Ruche looked down the pathway. At a table on a lawn, Max was sitting eating his breakfast.

'He's a lively boy – a real rebel. What about your wife? What is her name?'

'I don't have a wife.'

'You're a widower?'

'I never married.'

'Neither did I. It's strange, none of us married. Who is the boy, then, if he is not your grandson?'

'I love him like a grandson.'

'Did you ever try to do anything about his hearing?'

'His mother tried, but by the time she adopted him it was too late.'

'He said there were twins. Were they also adopted?'

'I don't like being interrogated. I won't say anything without a brief.'

Mr Ruche smiled – those were the first words they thought Sidney had spoken when Max brought him back from the flea market. He left Don Ottavio, and wheeled himself towards the table. Max didn't hear him coming, and turned round at the last minute. Mr Ruche quickly asked him if he had said anything about Grosrouvre, or about the Rainforest Library. Max shook his head and Mr Ruche asked him to say nothing.

'I promise. I've said too much already. It's my fault you're here. They didn't know about you, they only knew the name Liard. But when they brought me here I was so angry I said to Don Ottavio, "When Mr Ruche finds out you've kidnapped me, there'll be trouble!" When he heard your name, he jumped. He asked if your name was Pierre and I said yes, it was, and then he thought for a minute and said, "I think we'll have your Mr Ruche come here." That's when I knew I'd said something stupid.'

'No, Max, don't worry. We'll get out of this.'

'He said he knew you. Then he asked if you ever talked about Grosrouvre, and I said "Gros what?" like I didn't know, and he went away.'

'Good boy, Max.' He ruffled the boy's hair. 'Remember, don't say anything about Manaus or about the Library.'

'I'll play deaf and dumb!'

'No', said Mr Ruiche, 'if they try to force you, then tell them whatever they want, do you understand?'

Don Ottavio looked over at them and walked towards the table. 'Have you finished your little chat? You do know there are microphones everywhere, don't you?'

Mr Ruche's heart sank.

'Let the boy finish his breakfast. He's a growing lad, he needs to eat well. So, Pierre Ruche, shall we continue our stroll?'

They went off together.

'As I was saying, Elgar would never have risked leaving a physical trace of his proofs in case they fell into the wrong hands. But he might just have told someone about them.'

When he heard this, Mr Ruche shuddered, but Don Ottavio was so caught up in what he was saying that he didn't notice.

'But, if he had, the person he had confided in could have made them public there and then. Isn't that what you said about an expert? Therefore I deduced it couldn't be a physical object, so I began looking for a recording without a tape. Something that has a memory, but can't be played.'

Mr Ruche listened, but did not understand what he was talking about. Don Ottavio, clearly proud of his deductive powers, said triumphantly.

'A parrot.'

'You mean...My God? The parrot is the...'

'The what?'

Mr Ruche had almost said 'the loyal friend'.

'Yes, Pierre, the parrot.'

Mr Ruche couldn't believe what he was hearing. In a flash, he thought back over the months that he, Perrette and the children had spent studying the clues in Grosrouvre's letters trying to find out about the proofs when the solution had been right under their noses all the time. Sidney was Grosrouvre's faithful companion.

Back in the living room, Mr Ruche thought about Grosrouvre's letter. '"I've adopted a lot of animals since I've been here and, of course, we talk all the time. We've had many fascinating discussions." Grosrouvre told me everything I needed to know, but I wasn't listening.'

He looked up at Don Ottavio. It was clear from his face that he was quite serious. Don Ottavio caught his eye: 'What's the matter Pierre? You look confused.'

'I feel confused. You've just told me, in all seriousness, that Elgar confided his deepest secrets – and not just secrets, but mathematical proofs – to a parrot. I think I have a right to look confused. What am I supposed to say? "Elementary, my dear Watson"? You may have got used to the idea, but I haven't. Now I understand why you were so desperate to get the parrot back.'

As he said it, Mr Ruche realized that this in itself was another reason to believe Don Ottavio. There had to be a serious reason for a man in his position to spend so much time and energy kidnapping a parrot.

'If I want something, I don't like to have to wait', said Don Ottavio.

'It's ...it's unbelievable!'

'Unbelievable? You never saw the way Elgar talked to his Mamaguena.'

'His what?'

'Mamaguena. It was his name for the parrot you call Sidney.'

'It's a female, then?'

'Oh, yes. Grosrouvre bought her just after he got to Manaus – she was only a few weeks old – and for fifty years they were together. Wherever he went, she went. He took her on his trips into the forest and on river journeys when he was panning for gold. He talked to her for hours as though she was his oldest friend. You should have seen them together. When he was working all night in his library, she would sit on her perch and never say a word. I think he was more attached to her than to anything else. Except his proofs. And his library.'

There was a knock at the door, and the Tall Stocky Guy came in and said something quietly to Don Ottavio.

'Excuse me one moment', he said, leaving the room.

The interruption was a welcome relief for Mr Ruche, who was

having difficulty absorbing everything Don Ottavio was telling him. His first thought was how pleased Lea would be that the first parrot mathematician was female!

When Don Ottavio returned, Mr Ruche said quietly, 'You have everything you could want, then, Don Ottavio. You have the parrot. Keep the proofs, lock them up in a safe and leave us alone! Let the boy go and let us go home.'

'You will stay here for as long as I wish', said Don Ottavio coldly.

'No one speaks to me like that', shouted Mr Ruche, 'I'm not one of your hired henchmen.'

Surprised by the violence in Mr Ruche's voice, Don Ottavio fell silent. He felt as though he was seventeen and a poor Sicilian immigrant again, but his dark eyes glittered malevolently. In a moment, he was calm. Now he understood that though he could keep Mr Ruche here, by force, if necessary, it would be unwise to speak to him in that tone of voice.

Quietly, he said, 'The parrot hasn't spoken.'

'Sidney hasn't said anything?'

'Not a word.'

'I've always thought she was the most talkative parrot I'd ever met. She is a bit of a character', said Mr Ruche, secretly proud. 'Maybe she doesn't want to talk.'

'She *can't* talk', yelled Don Ottavio, 'she's lost her memory.'

Mr Ruche almost fell from his wheelchair laughing. Under all his threats and bluster, Tavio was ridiculous. He was still a boy playing at being a gangster.

'And here am I, Don Ottavio, like a *cretino*! Like some stupid little thief who finds a safe full of money and has no key, no combination and no way of breaking in. The proofs are still in that damned parrot's head. That's why I brought the boy. He is the only one who can help me.'

His eyes gleamed again. 'Do you know that, in the wild, parrots do not imitate the sounds they hear, or even the calls of other birds? Even in captivity, those who are kept with other parrots never learn to speak.' He stopped and thought. 'Why is it that they only learn to speak if they have human company?'

'Probably so that they can memorize mathematical proofs', offered Mr Ruche.

Don Ottavio's aviary was extraordinary. It was bigger than anything Max had ever seen. He stood outside talking to Sidney who was perched high up in the aviary. Sidney didn't respond. Head tucked under her wing, she clearly wasn't happy being a prisoner. Even though the aviary was huge, it was still a four-star prison.

Mr Ruche came and joined Max, a little out of breath. He told the boy what he had found out. Max watched carefully as Mr Ruche spoke, careful not to miss a word. When he had finished, Max called to Sidney, who flew down close to where Max was standing. Max slipped his hand through the bars and stroked the scar on the bird's forehead.

Max spoke softly to Mr Ruche. 'You know she hasn't eaten anything since she left Paris. If we don't do something, she'll die, I know she will. I don't care about anything else, I just want Sidney to be all right. I have to look after her, so...so I'm going to help them. If Sidney can give Don Ottavio the proofs, then let her.'

CHAPTER 24

Less is More

Archimedes: the siege of Syracuse; the beginnings of ballistics & a failed attempt

A limousine left the mansion at five o'clock with Don Ottavio at the wheel. Beside him, Mr Ruche watched as the countryside unfurled. The car swept past the Latomie del Paradiso and drove on through lush vegetation, past vast quarries and along the chalk cliffs. Don Ottavio was silent as they drove. The car swung left along the coast towards the Groticelli Cemetery. Tourists were out in force, marching in line like soldiers, their white legs exposed in baggy shorts, handkerchiefs on their heads. Don Ottavio slowed down. He sounded his horn and the tourists scattered like pigeons.

As they drove past, he said, 'I didn't quite tell you everything about my desire to own Grosrouvre's proofs. What I told you was true, but I didn't tell you the most important thing. The fact that Grosrouvre was studying maths was very important – if he had been working on anything else, this might not have happened.

'Have you ever seen a map of Sicily?' he asked Mr Ruche suddenly. He traced three lines on the windscreen, exactly as Max had done when they had studied Pythagoras back in Paris.

'In the ancient world they called it the Trinacrie, the three-pointed land. Cape Peloro in the north-east, Lilibeo in the west and Pachynum in the south-east. A triangle whose every side faces a different sea.'

He placed his finger on a point within the imaginary triangle. 'In the centre is the town of Enna, and from there mountain ranges run down to the sea, dividing the island into three sections. I was born on a triangular island – it was made for mathematics.'

Mr Ruche listened to Don Ottavio in silence.

'One afternoon, in my last year at school, my maths teacher drove me out here.'

He stopped the car, opened the window and pointed to a cave in the rock-face where, under a tangle of roots and branches, stood a ruin.

'We walked up to the cave, and my teacher knelt down and showed me the markings engraved in the stone. They had been almost completely worn away by time. He drew them out again on the ground for me. A sphere within a cylinder. The stone marked the grave of Archimedes.'

Don Ottavio closed the window and drove off. 'Why a sphere and a cylinder?' he asked. 'Because Archimedes had proved that the volume of a sphere was two-thirds that of the cylinder, and that the surface of the sphere was exactly four times that of the circle of the cylinder. Interesting, isn't it? I never got to go to the Sorbonne, like you. I was in the café opposite!' He laughed. 'Look!' As he drove, he got out his keyring.

'Watch out!' shouted Mr Ruche as they swerved to avoid a cyclist. Don Ottavio handed him the keyring. The fob was gold, studded with diamonds with a geometric figure engraved on it.'

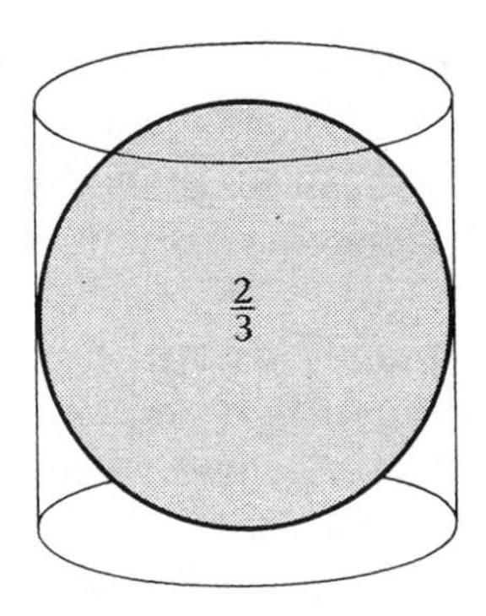

On the other side was the Sicilian coat of arms: a triangle with three legs running, one in each direction, linked by the head of a gorgon with serpents in its hair. The detail of the engraver's work was breathtaking.

'Do you understand now? Archimedes, the Trinacrie, Sicily. I've just thought...the three legs could be us – you, me and Grosrouvre.

Each running in different directions, but still connected.'

'I don't do much running nowadays', said Mr Ruche.

'I'm sorry, Pierre. But I never think of you as being...'

'... disabled?'

'You don't seem any different to me.'

'Three legs running – one of them is buried in Manaus, one of them hasn't moved for ten years, and you. I think you do enough running for all of us. If you're not careful, you'll wear yourself out.'

'I already have.'

'I still want to know why you brought us here. Why didn't you simply come to Paris?'

'I wanted you to see the mansion.'

'You brought Max and Sidney here before you even knew I was involved.'

'You really want to know why? I told you that my heart is weak. The doctor who examined you is a cardiologist. He's been treating me for years. He told me...well, anyway, I decided not to leave Sicily. I don't want to die in a strange land like my father. That is why I did not come to Paris.'

'So...none of us is running any more.'

They fell silent.

Don Ottavio accelerated as they crossed the Plain of Epipoleis. Mr Ruche opened his window and felt a warm breeze on his face. Don Ottavio stared straight ahead, his hair falling into his eyes. His air of authority had evaporated.

The car stopped near a ruined fortress. Don Ottavio got out and knocked on the door of a small cottage. A voice shouted through the door that the museum was closed. Don Ottavio knocked again. The door opened, and the caretaker recognized him at once. Without another word he went back inside and reappeared with a set of keys. He was clearly accustomed to Don Ottavio's visits.

'The fortress of Dionysius the Tyrant!' announced Don Ottavio proudly.

The fortress was impressive. It was surrounded by three trenches carved into the rock. In the third trench, Mr Ruche saw the remnants of the posts which had supported the drawbridge. In front of them five towers rose up, reddening in the sunset.

Mr Ruche looked away from the fortress. The car Mr Ruche had noticed following them had pulled up a short distance away. A man got out and stood beside it, looking through binoculars, trying to appear like a tourist – though the binoculars were more often trained on Don Ottavio than on the sea.

Leaning on his cane, Don Ottavio explained that the castle walls had run in a semicircle along the coast, enclosing the whole of the plateau. The walls to the north and south converged here, where the two men stood. Whether the enemy came from the mountains or from the sea, fifteen miles of ramparts protected Syracuse.

'Come over here!' said Don Ottavio, 'I'll show you something that will explain everything. Hurry, we must get there before it is dark.' Don Ottavio pushed the wheelchair across the uneven ground, ignoring any discomfort to Mr Ruche.

'Could you slow down?'

'We have to get there before sunset if you want to see the battle.'

He stopped the wheelchair at the apex of the fortress. To the east, the sea was already in darkness. 'I am sure that it was standing here that Archimedes saw for the first time that the surface of all liquids is curved', said Don Ottavio, 'curved like the surface of the earth, whether it is the sea, or just the coffee in a cup.'

Mr Ruche wasn't listening. He was admiring the scenery. Far off, the town glinted in the last of the sun's rays. It was magnificent.

'It was down there, on that peninsula, that the first Greeks arrived from Corinth. It was still an island then. To the right is the Porto Grande and to the left, the Porto Piccolo.

'The battle was fought between Marcellus, the greatest of all Roman generals, and Archimedes, the greatest of all Greek scientists. Syracuse was a rich and powerful island at the time and the most fertile in the Mediterranean. It was grain from Sicily that kept Rome alive.

'The battle took place in 215 BC. Marcellus attacked Syracuse by land and sea simultaneously.' Don Ottavio pointed to the Porto Piccolo with his cane. 'Sixty Roman galleys arrived in the port in battle formation, heading for the ramparts near the Porto Piccolo, where Archimedes lived. The archers immediately took up their positions on top of the walls. At the same moment, behind us' – Don Ottavio spun the wheelchair around to face the ramparts on the plateau – 'the Roman foot-soldiers came over the mountains and attacked the fortress, hoping to open a breach so they could invade the city.

'It was here that Archimedes' machines were waiting for them. He was not only a mathematician, but a formidable weapons designer. Using levers, boulders were catapulted over the walls like hail and crashed down on the Roman infantry. They had never experienced such powerful weaponry. The land attack was nipped in the bud, but the sea battle proved more difficult.'

Don Ottavio turned Mr Ruche's wheelchair to face the sea. He held the armrest so that he could lean on it. Don Ottavio was reliving the war. He recounted the story as though he had been one of those who defended Syracuse and had come back, two thousand years later, to tell his tale. From time to time as he spoke he pointed with his cane to places where the battle had raged.

Mr Ruche was caught up in the tale. For a moment he forgot about the kidnapping, Grosrouvre, everything that had brought him to Syracuse, and listened to Don Ottavio, captivated by his story.

'Suddenly, eight of the galleys broke away from the fleet. They were tied together in pairs by huge ropes, and on a great platform suspended between them was a terrible new weapon: a giant siege tower. These were terrifying contraptions: towers, several storeys high, with a system of ladders protected by an outer wall. End-to-end, they could easily reach the top of the ramparts. If they reached the walls, Syracuse was doomed. Soldiers armed and ready for combat waited at the foot of the ladders. Dozens of men pulled at the ropes to raise the towers while others put supports beneath them to keep them stable. The assault was imminent. Soldiers were already scaling the ladders when a huge rock sailed over the ramparts. Before it had hit its target, a second whistled through the air and a third. They hit the galleys, which tipped for a moment. The soldiers on the towers howled and fell, landing at the feet of the men below, who began to scream too. The siege towers crashed onto the platform between the galleys, sinking several of them.

'Rome's secret weapon, with which it hoped to invade Syracuse, had fallen to Archimedes' weaponry. The soldiers on the other galleys were disheartened by this setback, but still had absolute faith in their general, Marcellus. As night fell, they brought the galleys close to the walls of the fortress. Once there, they thought they were safe. Archimedes' weapons had a long firing range, making them ineffective while the galleys were sheltered by the walls – the rocks would simply fly over their heads – while their other weapons were useless. That, at least, was what Marcellus thought.

'At dawn, while the Romans were preparing to attack, Archimedes decided to use another of his inventions. Huge beams were pushed out across the ramparts and crashed down onto the Roman galleys, then hauled back using pulleys and ropes so that the attacks could be repeated.'

Don Ottavio began to recite: '"A huge beam swinging on a pivot projected from the wall and a strong chain hanging from the end had an iron grappling hook fastened to it. This was lowered onto the prow of a ship, and a heavy lead weight brought the other end of the beam to the ground, raising the prow into the air and making the vessel rest on its stern. Then, the weight being removed, the prow was suddenly dashed into the water as though it had fallen from the wall, to the great consternation of the sailors. The impact was so great that if the ship fell straight it shipped a considerable amount of water." That's from Livy's *History*.

'Marcellus gave the order for the galleys to disperse and place themselves around the ramparts so that Archimedes could not aim at them, but Archimedes had foreseen this. His catapults were as finely tuned as the pipes of an organ, each of them mounted over a different gate, and they could easily target individual ships.

'Marcellus ordered the fleet to keep moving, but the firing continued. The soldiers, though battle-hardened by fighting for many years under Marcellus, began to panic. They had never seen and certainly never suffered such an onslaught. The greatest Roman general was being defeated at Syracuse. He could not understand how such a thing could have happened.

'He would have understood if he had known what Archimedes had been working on for years: how long a lever should be, what counterweights were needed to fire the projectiles, everything there was to know about weights and balance. Using geometry, he had established the basic rules of mechanics. The people of

Syracuse were not surprised – they knew Archimedes all too well!'

Don Ottavio began to recite again: '"Sitting some way off, Archimedes worked, seemingly without effort, at a machine with many pulleys, drawing the galley towards him easily." In doing so, Archimedes completely destroyed one of Aristotle's great principles which had been accepted for centuries: the principle of impotence.'

'Of impotence?' asked Mr Ruche.

'Yes', said Don Ottavio, 'the name is mine. The principle is that if a force is weak and resistance great, then speed is nil. This is what Aristotle stated. If that's not a principle of impotence, I don't know what is! You'd agree that the force of Archimedes' pulling was weak? And the resistance of the galley on the sea was great? But the boat still moved towards the bank – so its velocity was not nil, therefore Aristotle's principle was completely wrong!

'The people of Syracuse were impressed by this new feat of Archimedes. They were familiar with his genius. Did you know that it was he who discovered that the royal crown had been tampered with, silver being added to the gold to make up the mass?'

Mr Ruche knew this story, but he listened with a sly smile on his face as Don Ottavio continued. In the end he couldn't help interrupting: 'It's amazing what you seem to get out of Archimedes. For half an hour now, you've been singing the praises of weights and balances when only this morning you were quite happy to bypass them. After all, what was Archimedes' work about if it wasn't balance and equality?'

Don Ottavio stared at Mr Ruche, stunned, but with something approaching admiration.

Mr Ruche went on, 'Here you are, the great Don Ottavio, trafficking in anything that comes to hand, but in awe of Archimedes, who spent his energies unmasking a forger. That's what surprises me.'

'You're right, I suppose', admitted Don Ottavio, 'nobody's perfect.'

'If you're not careful, you'll end up working for Interpol.'

'That's insulting! What I've just told you I heard for the first time on this very spot. My teacher explained it to me. You can't imagine the effect it had on me to know that someone from Syracuse had beaten the Romans at their own game. I was thrilled. Archimedes had thumbed his nose at all the snobs in Rome, all the northern Italians who came to Sicily to conquer it and treated us like dirt. He made me proud to be Sicilian! Two or three days after that, the same teacher told us in class about Archimedes' axiom. Do you know it?'

'No', said Mr Ruche, furious that after all his studying there was still something Don Ottavio could teach him about mathematics.

'Well, I'll explain it to you then', said Don Ottavio. 'The teacher told us it said – and I remember it word for word – "There is always a multiple of the smallest which is greater than the greatest". We didn't know what he was talking about, so he explained. "If you have a small object and a large one, you can always multiply the small object so that it is bigger than the large one." It was like a bomb going off inside my head. The bell rang for the end of class. I wanted to stay behind and talk to the teacher, but he was in a hurry. On my way home from school I sat down in one of the ruins we passed, and thought. It was the first time I had ever really thought about anything. I thought, "Tavio, you're the little piece", and everything suddenly seemed clear. What the teacher had said was that no matter how small you were, you could multiply yourself to be bigger than the biggest.

'The following Sunday, I passed a nobleman as I was walking in the village. My father bowed and tipped his hat, and I thought to myself, "You might be a count, but I'll be greater than you some day", and I felt warm inside. But how was I going to "multiply" myself? That's what I have been learning ever since: how to multiply myself so that I am better than the greatest. And I have succeeded.'

Mr Ruche said nothing. Then, because what Don Ottavio had said troubled him, he said, almost to himself: 'There are always small people...and some of them will want to be greater than the greatest. And you became great.'

'Yes, but I have never forgotten that I was small, so I keep on making sure I am multiplying.'

'I understand', said Mr Ruche. '"Give me a fixed point on which to stand and I will move the earth." Archimedes said that. The smallest thing can lift the most enormous weight with levers.'

'Let me tell you how the battle ended', said Don Ottavio. 'On this spot, the greatest Roman general was beaten by the greatest Greek mathematician. But instead of retreating north, Marcellus used a coward's tactics: he laid siege to the city. What he couldn't take by force, he intended to take by starvation. Two years later, Syracuse was still holding out.

'In the end, it was the length of its own ramparts that defeated Syracuse. In a siege that lasted more than two years, how could they watch every inch of them all the time? One night, when the people of Syracuse were feasting, a bunch of traitors opened one of the gates to the city. The Romans surged in, and the city was taken.

'Marcellus rushed in to see the machines that had defeated him. He was amazed. Now he understood why he could not win, faced with such an adversary, why he would never have taken the city if it hadn't been for traitors. He went to find Archimedes.'

As Mr Ruche watched the last glow of the sunset over the city, and Don Ottavio recounted the night in 212 BC when Syracuse fell to the Roman army, Mr Ruche imagined the pillage and the desecration and the fires raging through the city. Drunken soldiers singing in the streets, carrying the spoils they had looted from the homes of the rich. When dawn broke the following morning, it was over a ravaged city.

'Archimedes sat at the foot of the ramparts. The tide had not yet washed away the shapes he had been drawing in the wet sand. There was a bloodstain on his white toga. Caught up in his geometry, Archimedes had not heard footsteps approach; he had not turned. The soldier, angry not to find anything of value, trampled Archimedes' diagrams in disgust and then killed the great man with his sword.'

Don Ottavio was silent for a moment, then he said, 'In a few hours, my teacher had made me proud to have been born here, and had taught me that that itself was not enough. He taught me the pain of defeat and the desire for revenge. I suppose I grew old that day. Archimedes was seventy-five when he died.'

Don Ottavio felt strangely moved. He was not an emotional man. He surrounded himself with advisors and bodyguards, with lawyers and bankers. He had never opened himself to someone as he had just done – except perhaps to Grosrouvre. And even with

Grosrouvre, he had not been so frank or so emotional. It was the landscape, the city, which had made it possible.

'It's getting late', he said. 'We should go back.'

'My God', thought Mr Ruche, 'I've forgotten to phone Perrette. I told her I would phone her every evening before eight o'clock. She'll be worried sick.'

Don Ottavio pushed the wheelchair slowly back along the plateau with difficulty. Mr Ruche could hear the man's laboured breathing. They reached the limousine, and Don Ottavio helped him into the comfortable leather seat. The car pulled away soundlessly. Behind them, Don Ottavio's bodyguards followed in the other car. Slowly, they made their way back to the mansion.

Mr Ruche was thinking about Hippasus of Metapontum. Like Don Ottavio, he had started life a poor boy and ended it a rich man. He had made his fortune in Sicily, and no one knew how he had come by the money. Hippasus was not interested in theory and would use any method or any ruse as a means to an end, just like Don Ottavio.

'Forty-four thousand nine hundred and sixty three billion, five hundred and forty...'

Mr Ruche was shaken from his thoughts, wondering if Don Ottavio was trying to impress him with his bank balance.

'...million years. That's how long it would have taken Archimedes to lift the world using a lever. Some mad Englishman worked it out', said Don Ottavio, laughing. 'Not that it matters – all that matters is that it could be done.'

'His love of Archimedes has a strange effect on Don Ottavio', thought Mr Ruche. 'It makes him think like a mathematician. In maths, time doesn't matter. Archimedes could have lifted the earth using a lever, that's all that matters.'

'The tomb I showed you earlier, the one my teacher showed me, isn't Archimedes' tomb, but what difference does it make? I love legends but, as you've seen, I'm more interested in the real world.'

Since Mr Ruche had left for Syracuse, Perrette had been thinking. She was certain that Grosrouvre must have sent Mr Ruche something, some little clue about his proofs. She had been spending her days ferreting around in the Rainforest Library.

On her first visit, she had found everything exactly as Mr Ruche had left it. Suddenly, she remembered the new alarm. She only had forty seconds to disable it! She started to type in the code, but halfway through she forgot. Then she remembered the phrase Mr Ruche had taught her to help her remember the number:

'How I want a drink, alcoholic of course, after the heavy lectures involving quantum mechanics.'

It was the number of letters in each word: How = 3, I = 1, want = 4, a = 1, drink = 5, etc. She cancelled and began again, typing in π to the first fifteen decimal places: 3.14159265358979. Thirty-five seconds – it was a close call.

She opened one of the crates of mathematical journals, and found two neat packages tied with string. She cut the strings and started to transfer the journals to the last empty shelf, careful to keep them in order. She wondered whether Mr Ruche had left them in the crate because they were recent. Most of them were American, but some were French and Russian. Perrette looked at the titles, but found it difficult to work out why they had been separated into two piles. She leafed through the first of them and noticed on the contents page that one of the articles was underlined.

'Mother!' Lea was calling her from the balcony. 'Phone! It's from Syracuse!'

It was Max. He was speaking to Jonathan, and Mr Ruche was telling Max what Jonathan said. Each of them spoke to him in turn. When Lea hung up, Perrette broke down in tears. Jon-and-Lea didn't know what to do. They had never seen their mother cry before.

When she got back to the Rainforest Library, she returned to the piles of journals. On the contents page of each, one article was underlined. In Number 44 of *Inventiones Mathematicae*, 1978, the underlined article was Barry Mazur's, 'Rational isogenies of prime degree'. In Number 29 of *Communications on Pure and Applied Mathematics*, 1976, the article was by Goro Shimura, 'The special values of the zeta function associated with cusp forms'. As she turned the pages, she came upon the opening lines of Goro Shimura's article:

1. Introduction

For a positive integer k and a Dirichlet character χ modulo a positive integer n such that $\chi(-1) = (-1)^{\chi}$, let $G_k(N, \chi)$ denote the vector space of all holomorphic modular forms $f(x)$ satisfying

$$f(\gamma(z)) = \chi(d)(cz + d)^k f(z) \quad \text{for all } \gamma = (\begin{smallmatrix} a & b \\ c & d \end{smallmatrix}) \in \gamma_0(N)$$

where z is the variable on the upper half-plane, $\gamma(z) = \dfrac{(az + b)}{(cz + d)}$, and

$$\gamma_0(N) = \left\{ (\begin{smallmatrix} a & b \\ c & d \end{smallmatrix}) \in SL_2(Z) \mid c \equiv 0 \pmod{N} \right\}$$

Suddenly she felt very tired.

Mr Ruche was wide awake. He was trying to digest everything he had learned recently. He now knew the identity of Grosrouvre's 'faithful friend', and of the gang who wanted his proofs. It was strange to think that Perrette had anticipated Tavio's involvement. As for the 'faithful friend', Mr Ruche had to laugh – the answer had been staring them in the face for seven months. As recently as a week ago, Mr Ruche was certain he would have to go to Manaus to find the 'faithful friend'!

Suddenly, he remembered something that hadn't seemed important at the time. There were no books by Archimedes in the Rainforest Library. Don Ottavio's library, however, boasted Archimedes' complete works.

Mr Ruche took down the first book that came to hand and opened it. It was Plutarch's *Life of Marcellus*, illustrated with wonderful miniatures by Girolamo of Cremona. Mr Ruche looked at the publication date – MCDLXXVIII. He whistled softly. This was one of the first books ever printed!

There were other books by philosophers and historians recounting the life of Archimedes: Livy, Cicero and Polybeus. He took down a copy of Archimedes' *On Floating Bodies* and found the passage that Don Ottavio had mentioned earlier: 'The surface of any liquid at rest has the form of a sphere whose centre is the centre of the earth.'

There was a noise, and Don Ottavio appeared at the door.

'Can't you sleep? I noticed your light was on.'

'So you came in, just like in a forties film. Come in', said Mr Ruche.

'Shhh', said Don Ottavio, nodding to the bed where Max was asleep.

'He's got a nerve', thought Mr Ruche. 'He kidnaps Max, puts him on a plane and takes him a thousand miles away and now he's telling me to be quiet in case I wake him up!'

'Max is deaf. You can talk as loudly as you like', he said.

'You were looking at the books? Wonderful, aren't they?'

To Mr Ruche's surprise, he reeled off the titles from memory: '*Quadrature of the Parabola, On the Sphere and the Cylinder, Measurement of a Circle, On Conoids and Spheroids, On Spirals, On Floating Bodies, The Method, The Sand-Reckoner.*'

He put his glasses on and took down a volume.

'*The Sand-Reckoner.*' He began to recite: '"King Gelon, no one believes that the number of grains of sand is infinite and, not only the sands of Syracuse, but those which surround all lands inhabited and uninhabitable."' He shot a glance at Mr Ruche as if to say, 'I may wear glasses, but my memory is perfect. Is yours?' He became passionate now. 'Here, Archimedes goes to work. With the smallest things in the world – grains of sand – he will measure the greatest, the universe itself! Do you know how many grains of sand there are? A number sixty-four digits long. One evening in Manaus, Grosrouvre told me how Archimedes went about it. It took hours to tell, but he was a great storyteller. He always thought of maths as stories. The more numbers there were, the more we drank, and we were a little drunk by the time we finished. He told me that Archimedes had devised a system which could handle numbers up to...' – he adjusted his glasses and thumbed through the book – '...80 million billion digits. I dream about that number sometimes. A myriad of myriads to the myriad power. Wait', he stopped suddenly, 'that night he told me that when you were students, you liked Thales and he preferred Pythagoras. I always preferred Archimedes.'

Gesturing towards the books, Don Ottavio said, with a quiver in his voice, 'This is all I have left of Elgar. He gave them to me years ago. They used to be in his library. I don't think I told you about his library.'

This was a dangerous moment, thought Mr Ruche. He had to be careful not to slip up.

'It was one of the most beautiful libraries in the world. It contained only books on mathematics, many of them very rare. He built it up himself, book by book – it took him years and cost him a fortune. Everything he earned he spent on his library.

'I helped him whenever I could. Sometimes, if he was short of money, I would make up the difference; sometimes I applied a little pressure to people who were reluctant to sell. I don't know much about books, but you work with them all the time. You would have loved the library. The strangest thing was that such a thing was there, in the middle of the jungle. It seemed ironic to have all those books full of calculations and theorems in the middle of nowhere, but that was Elgar for you!

'He was very careful, though. He kept the books in a cool, dry room because, in the rainforest, humidity destroys everything. He ordered machines to control the humidity and a chart-recorder – you know, the machine you sometimes see in hospitals that traces a line of a roll of paper. It broke down once when I was there. I never saw him in such a state. He was so obsessed with his library. Books, well, they were never...'

'...your cup of tea?' suggested Mr Ruche ironically.

'And it was all burned.'

'Burned?' said Mr Ruche, feigning horror.

'When his house burned down, the library was destroyed and Elgar was killed.'

Mr Ruche was angry now, but he had to be careful not to give himself away. Nothing he said should give Don Ottavio any idea that he already knew the story all too well. He could still hear the words of Grosrouvre's letter.

'It reminds me of something – something that happened not far from here, in Crotona. Grosrouvre probably told you about it. There was a rich man living there, Cylon, who admired the work of Pythagoras' school and wanted to be admitted to the ranks of his mathematicians. The Pythagoreans found him a little odd, and refused. Cylon was furious – he wasn't in the habit of being refused something he wanted. One night, when the school was in session, he and his cronies set fire to their house. All the Pythagoreans died in the fire, except one.'

Don Ottavio went white. He stood up and, for a while, he said nothing, steadying himself on his cane.

'What rich, powerful man had you in mind, Pierre? Are you suggesting I set fire to Grosrouvre's house? Are you accusing me of killing him?'

Mr Ruche was frightened. Don Ottavio's rage was terrifying.

'You think I would do something like that? Kill a friend...'

'...a friend who had refused to give you what you wanted. He was probably the only person ever to refuse you anything.'

'It's true, Elgar did refuse me, and he was the only one who had ever done so. Yes, I was angry. But he was due to give me his final answer that night.'

Mr Ruche bit his lip. He knew all of this already. 'They'll be back tonight. Believe me, πR, I won't give them the proofs. I will burn them as soon as I have finished this letter...'

'When my men got there, the house was already ablaze. I arrived shortly after. It was terrible. It was a huge wooden house and there was no way to put out the fire. There was nothing we could do to save Elgar. I felt dreadful. We left quickly. The police would have arrived soon, and it was best that we weren't seen there.

Don Ottavio leaned down and looked Mr Ruche in the eye: 'You have to believe me, Pierre, I need you to believe me. That was one of the reasons I asked you to come here.'

'You didn't need to kidnap my grandson to convince me. Do you really not know what Elgar's answer would have been?'

The volume of Plutarch was still open on the desk where Don Ottavio had laid it. Girolamo of Cremona's miniatures danced on the page. Don Ottavio stared at them, talking to himself. 'It would have been like sharing one of the secrets of Archimedes.' He lifted his head quickly, his white hair glinting in the lamplight.

'Listen to me, Pierre, even if I had never been friends with Elgar, I had no interest in seeing him dead. His death was a disaster for me, because his proofs died with him.'

'Maybe you thought you could force him to tell you', said Mr

Ruche, 'then you would have been forced to kill him, because he could have published them any time he liked.'

'He would never have published them. He would have preferred that he and I were the only ones to have them than to make them public. That was what I wanted – not to take them from him, but to share them with him. Just the two of us.'

He thought for a moment, and when he spoke again he was calm and in control. 'The facts are that Elgar is dead and I still do not have the proofs. Those are facts.'

Mr Ruche decided it was time to speak up. 'It's still a fact that the fire happened just before you were supposed to meet him, just before he was supposed to respond to your ultimatum, and it was the fire that killed him. You can't deny it. Whether the fire was an accident, or whether Grosrouvre committed suicide to avoid telling you, he would have burned his papers so that you couldn't get at them. You are responsible for his death. You refused to respect what he wanted, because what you want always comes first. You didn't respect his wishes. You didn't behave like a friend.'

Don Ottavio sat down. Mr Ruche's last phrase had hurt him.

'I have something else I'd like to say', Mr Ruche continued. 'It's about what you told me this afternoon, about your teacher and Archimedes. I was very touched. I have a feeling you've never talked like that to anyone before. I can understand why you needed to rebel, why you needed to be proud of yourself. And, thanks to your teacher and to Archimedes, you found a way. But what you've done and the methods you've used haven't changed the world, Tavio. You've had your revenge, but it hasn't made anything better – it's just made the world a little worse. Your money is like a poison. All right, you've become Don Ottavio, people raise their hats to you. You live up here in the big house. People are afraid of you. But out there, kids as young as Max are hooked on heroin. It drips into their veins like a poison, and in the end it kills them.'

'How dare you say that! I have never had anything to do with drugs, never!' shouted Don Ottavio. 'I have my standards too, Pierre, they're simply not the same as yours.'

'It doesn't matter. You have survived by trampling over others. You had Max – a child – kidnapped just to get what you wanted.'

'Don't forget the parrot.'

'A kid and a parrot. And another thing: Archimedes' axiom, the thing that gave you your strength to be greater. I read something about it just before you came in.' He leafed through the book. 'Here it is: "any segment, no matter how small, if it is halved successively, can be made smaller than any other segment." '

Don Ottavio's concentration was obvious as he tried to understand, but his eyes shone, as they always did when Archimedes was mentioned.

'Which means that you can be "divided" just as you were "multiplied" and made as small as anyone else', finished Mr Ruche, coldly.

When Don Ottavio had left, Mr Ruche went over to Max's bed. The boy was sleeping like a log. This was the first time Mr Ruche had slept in the same room as one of Perrette's children, in fact it had been years since anyone else had slept in his room. The sound of Max's breathing, deep and regular, disturbed him. But today, he had realized something which was priceless – he loved the boy. In the park, he had heard himself say, 'He is like my grandson', and only moments ago he had referred to Max as 'my grandson'!

Mr Ruche wheeled himself to the balcony. Despite his predicament he loved this area – the warmth and the mingled scent of flowers rising from the gardens. The moon was full and flashed on the sea where the terrible battles Don Ottavio described had been fought. Lights moving in the park caught his eye. The security guards were making their rounds carrying powerful torches, guard dogs walking beside them.

The lights shook him from his thoughts of ancient battles, he had completely forgotten that he was still a prisoner here in this beautiful but tightly guarded castle. In fact, it was subtler than that – he was not exactly a prisoner but neither could he simply leave. He shivered. It was beginning to cool down and he wheeled himself inside.

The sun was already high in the sky when the gardener opened the great gates to let Max into the aviary. High above, Sidney was perched on a thatched roof. Max called softly to her, and she ruffled her feathers and flew down to perch on Max's shoulder. Mr

Ruche, who was watching from afar, recalled Plato's definition of a mathematician: 'a birdman in an aviary capturing birds of brilliant colours'.

Max walked out of the aviary with Sidney, into the dazzling sunlight. Now that she was free again, Sidney immediately ended her hunger strike and began eating the seeds in Max's hand.

The big day had arrived. Don Ottavio felt he had everything on his side. He had spoken to Max and was convinced that the boy would help him. For Max, the only thing that mattered was that Sidney was free again.

They all entered the mansion. As they crossed the great hall, they stopped by a padded door. Don Ottavio opened it. Max and Sidney went in, but when Mr Ruche tried to follow Don Ottavio stopped him, saying, 'The fewer people who hear the proofs, the better for everyone.'

Mr Ruche had to agree.

Inside the room, a series of reel-to-reel tape recorders were connected to an impressive mixing desk, with dozens of sliders and lights. There was a projector. The walls were soundproofed and the floor was carpeted. In the middle of the room hung a microphone, with a perch and a feeding tray in front of it. Max placed Sidney on the perch and sat on a chair facing the microphone. Don Ottavio sat at the mixing desk. There was no sound engineer. Don Ottavio had decided not to speak to Sidney directly, Max would say everything. Don Ottavio had given Max a notebook with a list of simple words chosen for their emotive value – or so Don Ottavio hoped. He had consulted a number of experts who advised him that for Sidney to regain her memory she would need to be reminded of the world she had been part of before she received the blow to her head. The words were bait to draw Sidney's story from her.

Don Ottavio pressed a button, and a red light came on over the door of the studio. Mr Ruche knew that the session had begun. He sincerely hoped that Sidney would regain her memory. That would be the end of the story. It was true that the bastards would have won, but he could do nothing to fight someone with so much power.

However, equally, Mr Ruche knew that if Sidney talked, Don Ottavio would never let her go. He might even have the parrot put to sleep. Mr Ruche was disgusted by this, and began to change his mind and hope that Sidney would not get her memory back. For as long as she was an amnesiac, she was safe – albeit a prisoner. It was like squaring the circle: whichever way you turned, you were trapped.

Don Ottavio nodded to Max, who began to read out the list. He read the first word, waiting for Sidney to react, then repeated it, speaking to her gently all the time. Sidney remained silent. After each new word, Max talked to her, coaxing her to remember.

Don Ottavio followed the conversation through his headphones. Each time a word failed to get a reaction, he looked disappointed. The experts had warned him that he would have to be patient, that it was impossible to tell when the parrot's memory might return. He felt angry and powerless. There was no mirror here that he could twist to open the secret doorway. The proofs were hidden somewhere much more inaccessible than his museum of stolen treasures.

'Elgar', 'Manaus',...the list continued. When Max came to the last word, he read it to himself. This was the word Don Ottavio had pinned his hopes on. Max looked quizzically at Don Ottavio, who nodded. Max did not know what it meant, he simply said it softly: 'Mamaguena.' Don Ottavio held his breath.

Sidney looked at Max, who repeated the name. Sidney could not remember ever having been called Mamaguena. It was as though she had been born at the flea market nine months before. Memories of the fifty years she had spent in Manaus had been wiped clean, like a damaged computer disk. She was deeply amnesiac. She might never regain her memory. Don Ottavio turned pale.

The light went out, the projector hummed and a picture of a large wooden house in the middle of a forest appeared on the screen. In front of the house, a man stood facing the camera. He was about seventy, tall, with dark hair, and wore a white linen jacket and trousers. His shirt was open at the neck, and he looked strong and healthy. This was Grosrouvre in front of his house in Manaus. The film was silent. Sidney didn't blink.

The light came on again.

Don Ottavio took off his headphones. Sidney gulped down some water and nibbled at the seeds. Max did not know whether to be happy or sad.

In the corridor, the light above the door went out.

'You can't leave without seeing the sea.' Don Ottavio helped Mr Ruche into the limousine.

'Are we going back to Paris?' asked Mr Ruche.

'There's nothing left for you to do here. This morning's experiment was conclusive: the parrot is not going to get its memory back. There's no point in persisting.'

Mr Ruche breathed a sigh of relief and settled himself into the comfortable leather seat. He was getting used to this luxury. The limousine took the coast road lined with citrus and eucalyptus trees. It was cooler here.

'See that, in the water? That's not reeds, it's papyrus!', poured out Don Ottavio.

'Can we stop? I'd like to pick one.'

'I'm afraid it's illegal.'

'I can't believe a man who kidnapped a child is telling me I can't pick a single stem of papyrus because it's breaking the law', Mr Ruche laughed.

'This is the only place in Europe where it still grows wild', said Don Ottavio. 'In Egypt there isn't a single plant left. It didn't last as well as the pyramids. It probably won't last much longer here – the water is too salty and the roots are exposed to the air. To survive, it has to be in deep water. All Archimedes' work was written on papyrus, but none of it was found. The only versions of his manuscripts were copies on paper or parchment.'

The limousine headed towards the northern coast of Sicily. The route ran along the sea for several miles. There were beaches, not soft, sandy beaches but dozens of bays of pebbled beaches where the cliffs dropped steeply into the sea.

Don Ottavio's voice shook Mr Ruche from his thoughts, 'I have a proposition to put to you. We'll all go to Manaus: you, me and the boy.'

Mr Ruche shuddered. 'You're mad! I'm not going to Manaus, I just want to be left in peace. Anyway, Perrette would probably call the police. She's held off until now, but...'

Don Ottavio's face became a mask, and he said coldly, 'She wouldn't be very wise to do that. It's all gone very well until now.'

'You think so?'

'Tell her not to do anything stupid. It will soon be over.'

'Why should we go there', asked Mr Ruche, realizing that Don Ottavio had already made his decision. 'Can't you see the parrot doesn't remember anything?'

'The experts said that she should be in the same surroundings, in the place where she was before she lost her memory.'

'But the house has burned down. There's nothing left.'

'She lived in the forest near Manaus for fifty years, on the banks of the river. Even if the house is gone, it is closer to where she's lived most of her life than this mansion, or your bookshop in Paris. I give you my word, if the parrot doesn't talk when we get there, the three of you can go home and I'll leave you in peace.'

'And if I won't go?'

'Then I'll keep the parrot here. And if I keep the parrot, Max won't want to leave.'

'You're repulsive!' Mr Ruche didn't know what else to say. 'You have no right to keep that parrot.'

'Are you saying she's yours? Where did you buy her? Where are her papers? You have nothing at all to link you to the parrot, Pierre, and you know it.'

He was trapped. Don Ottavio had thought of everything. He wanted to scream.

'On the other hand', said Don Ottavio, 'I can prove she is mine. I have all the necessary papers.' He pulled over at the side of the road and took a leather wallet from the glove compartment. In it were a number of papers, which had been stamped. Everything looked official. Don Ottavio moved to put them back, but Mr Ruche stopped him. He examined the certificates but, as far as he could tell, they were genuine documents from the customs office in Palermo.

'Someone in my position can't afford to break the law', said Don Ottavio sarcastically.

He started the limousine and drove off. Mr Ruche decided that he had no choice. They would all have to go to Manaus.

'Look!' Don Ottavio pointed out to sea. 'That's the Rock of the Two Brothers.' He was silent for a moment. 'The boy was happy this morning. He obviously respects you and he's very fond of you. You're lucky.'

Mr Ruche couldn't stop himself from retorting. 'It's not like paintings, or even proofs. It's something that can't be bought. You have to earn it.'

'I've decided to take care of his future', said Don Ottavio. 'I'm going to leave him something in my will.'

'*You've* decided? Who are you to make decisions about us?'

'Not about you, about the boy.'

'We don't want your money – we don't need it.'

'You can't stop me from leaving him something.'

'No, but you can't force us to accept your money.'

Don Ottavio was about to say, 'It's none of your business, you're not part of his family', but kept quiet. 'No one can decide until the boy comes of age. Then he can decide for himself. Maybe by then...after all, medicine is making progress all the time. What right have you to take away a chance like that?'

Albert was sitting at the only table in the bar, drinking a glass of Marsala. It was not his first glass. The Tall Stocky Guy sat down. Albert barely looked up.

In a thick Italian accent, the Tall Stocky Guy asked, 'What brand of cigarettes are those?'

'None of your business.' Albert looked threatening, but a little dazed. He looked at the man who had spoken. He was big and stocky.

'Aldo', said the Tall Stocky Guy to the barman, 'another Marsala for Mr...'

'Albert.'

'Is the Renault 404 yours? It's a beautiful car. You've come a long way', said the Tall Stocky Guy.

'She's a good car', said Albert, 'never let me down yet.' He stubbed out his cigarette, took out his packet and offered one to the Tall Stocky Guy.

'I don't smoke.'

'Then why did you ask me which cigarettes I was smoking?'

'I have a photo of you with a cigarette hanging out of the corner of your mouth and I wanted to know what it was. Now I know it's a Gitane.'

'What photo?'

'This one.' The Tall Stocky Guy took out the photo of the Louvre pyramid from the Japanese magazine.

Albert looked at it, amazed. 'Where did you get this? I've never seen that photo before...' He thought for a minute. 'Although I do remember when it was taken...'

The Tall Stocky Guy leaned over and whispered conspiratorially: 'It was thanks to your cigarette that I was able to trace the parrot and find the kid.'

Albert shot out of his chair. 'What d'you mean?'

'One morning at Roissy Airport you refused to take a passenger from Tokyo...I was the passenger and I recognized you. Same cap, same cigarette in the corner of your mouth. You're famous for it! The man at the registry of taxis recognized the description immediately.'

'Shit!' Albert sank back into his chair.

'Aldo, get the man another Marsala.'

Albert drained the glass in one gulp. He was to blame for everything.

'Here comes your friend', said the Tall Stocky Guy. The limousine pulled up in front of the bar. Albert stood up, he could see Mr Ruche through the windscreen. He ran out to the car.

Before he could say anything, Mr Ruche reassured him: 'Everything's fine, Albert. We're going to go to Manaus for a couple of days. You go back to Paris. Tell Perrette not to worry, everything will be all right. Tell her I'll call her.'

'What about Max?'

'He's fine. And be careful driving back, Albert.'

As he got behind the wheel of the Renault, Albert spotted the name of the square where he was parked; Archimedes Piazza.

CHAPTER 25

Clear Blue Sky

High above the rainforest; a solitary bird & a death in the afternoon

Max closed his eyes. His face was tense – he hated flying. The pressure on his eardrums was very painful. He took long, deep breaths the way Perrette had taught him, and began to feel calmer as Don Ottavio's private jet began to climb.

Sidney's feathers stood up like hackles, and she held on tightly to her perch, which had been securely fastened to Max's armrest. The Tall Stocky Guy sat behind Max, stretched across two seats. From here, he could keep an eye on Sidney. On the other side, Don Ottavio and Mr Ruche were talking about probability and the difference between 'improbable' and 'impossible'. What was the probability that a parrot from Manaus belonging to an old man should end up in a bookshop in Montmartre owned by an old friend he had not seen for fifty years? Why had Max gone to the flea market that morning, the very morning that Sidney was there? How had the boy and the parrot come to be there at exactly the same moment? They could trace the reasons for each of these, but it didn't explain anything. The probability that it might happen must be infinitesimal, but not zero. It was totally improbable, but not impossible. Don Ottavio held the final piece of the puzzle and, as the stewardess served the meal, Don Ottavio explained everything he knew to Mr Ruche.

'I told you there were traffickers in rare birds interested in the parrot. This is where they come in. After the fire, when I had worked out that Grosrouvre had probably explained his proofs to the parrot, we tried to find it. It had disappeared. It had escaped from the burning house and flew to the bar where Grosrouvre usually drank, where it talked and talked. They couldn't shut it up.

Nobody could understand what it was talking about. The traffickers were sitting at one of the tables. They often come to Amazonia to get rare birds. They knew they could get a good price for the parrot, so they caught it. By the time we found out, it was too late – they had left Manaus. It didn't take long to find out they had gone to Paris – there's a lot of trafficking through Paris. I sent two of my men to find it. They took care of the traffickers and got the bird. That would have been the end of it, but one of my men' – he pointed to the Short Stocky Guy – 'let the bird escape. I could have...but he's an excellent marksman. It's the only thing he has going for him, but in our business, it's pretty useful. Where was I? Oh, yes...the parrot escaped at the flea market and they chased it all over the place. They were about to catch it when your grandson appeared...well, you know the rest.'

As the plane flew over the Atlantic, Perrette, thousands of miles to the east, was just opening the newspaper. As she scanned the front page, she spotted something and exclaimed, 'It can't be!'

They spent the first night in a splendid turn-of-the-century palace in Manaus. The newspapers all carried the same story about the disappearance of the Blue Ara. Don Ottavio showed the story to Mr Ruche who passed it to Max:

STILL NO TRACE OF THE LITTLE BLUE ARA

The search for the missing little blue ara has made little progress, it was reported yesterday. Its disappearance follows an attempt last year to mate the rare bird with a female, bred in captivity, which had been set free.

The spix ara, or little blue ara, is the rarest of parrots, with only one known specimen in the wild. The experts who first spotted it have been watching it closely for some years. The parrot had begun to try to breed with other species of parrots in the area. In order to ensure the survival of the species, seventeen females raised in captivity were brought together. Before the first of them was set free, it had to undergo a rigorous re-education to ensure that it could find its own food and live in the wild.

Unfortunately, it would seem that the companion the experts had chosen made little impression on the wild bird, which appears to have moved to another part of the Brazilian rainforest, possibly with a female macarena.

Since then, there have been no sightings of the parrot. The female, which was to have been his mate, has been returned to an aviary.

Early the following morning, they left for Grosrouvre's estate. It was on the banks of the river in a forest clearing. The extraordinary house that Max had seen in the short film was a ruin. Only one wing of the building remained standing. A tribe of Indians lived there now.

Two large mobile homes were parked on the riverbank. This was where they would stay. Don Ottavio wanted to begin Sidney's interrogation immediately, confident that this time he would succeed. The parrot was back in the world in which she had lived for fifty years. It was now or never. Max began to read the list – the words were not the same as those he had read in Syracuse.

Mr Ruche was tired. He was getting ready to wheel his chair into the van when an Indian woman of about fifty came over to him. She introduced herself as Melissa. She had been Grosrouvre's friend for many years.

'You are Mr Elgar's friend, the friend from Paris? He tell me much about you.' She looked at the wheelchair. 'He did not say about your legs.' She sat on the ground, folding her dress beneath her. She was looking into the distance as she spoke. 'When Mr Elgar come to our village, I was little girl. I see this man, this giant with beard in our village. He work the rubber, it is difficult work, all day in the forest, but Mr Elgar was never tired. The other men do not like Indians, they are bad to us. He was not like the others. He did not threaten us or take things from us. He could have', she said, clearly proud of him. 'When he want something, he pay for it.

'He come here many times and then he stay; he is like us. He is poor like us. I grow up, I watch him. His head was...' she looked up to the clouds above. 'He always write on pieces of paper and put the paper in his pockets. Writing make him feel better. The wise woman say that is his medicine.

'One day, he say to me "I am going to the gold river, I will go to find gold and diamonds". He was a *garimpeiro*. I did not see him again for many years. I am a young woman. My parents say that I marry but I say no.

'One day, he come out of the forest. I do not recognize him. He has no beard and he seem taller than before. I go with him to Manaus. He has much money. He buy many, many books. I like being with him. He start to think about something all the time. All night he write in his room. In the morning, he sleep. Mamaguena is always with him. I get jealous.'

Melissa talked for a long time. She said that when Grosrouvre disappeared, she had not returned to her village because of the daughter. 'When she has husband I will go back. This is my daughter.'

A young, dark-skinned woman was walking towards the road. She was about twenty, tall and slim.

'Sorbonne!' called Melissa.

The girl signalled that she was in a hurry and walked on.

'What did you say her name was?' asked Mr Ruche.

'Sorbonne.'

Mr Ruche looked surprised, and she explained. 'Mr Elgar say all the time, "Sorbonne was very beautiful", so when I have my daughter, I call her Sorbonne, so she is beautiful.'

Mr Ruche laughed, though he was more touched than he wanted her to know. He watched the young girl walk towards a beat-up car that was waiting by the road. He went into his mobile home. It was air conditioned and had every comfort possible. He lay down on the soft bed and fell asleep.

He woke up to find someone shaking him by the shoulder. It was the Short Stocky Guy. 'Come quickly, Don Ottavio wants to speak to you. He's not well.' He helped him to the other caravan, and then left the two men together. Don Ottavio was lying on his bed, his face livid.

'Pierre, it is important that you believe me when I say I did not burn the house, I did not kill Elgar. Yes, when he refused to give me the proofs I was angry, very angry. Why should he give them to a parrot and not to me? I don't know what happened. An accident. Everything was burned.'

He stopped and tried to catch his breath. 'Do you think he did it himself?' He brought his hand up to his chest.

'You need a doctor.'

'No, Pierre. There comes a time when you can't multiply yourself any more. Time to stop. I knew I should not have left Sicily. I'll die like my father did, far from home. Things always turn out the way you feared they would.'

'I want to tell you something', said Mr Ruche leaning towards Don Ottavio. 'Not long ago, Elgar did get in touch with me.'

'You think I didn't know? I investigated everything the moment I knew you were involved. I knew he had sent you his library.'

Mr Ruche looked at him, startled, then he blushed.

'You're a pretty good liar, Pierre. Did philosophy teach you that? I thought it was supposed to teach you the truth.' He stopped, clearly exhausted. 'Take care of the library – it is all that is left of him. I don't think the parrot will speak now.'

There was a gunshot nearby. Mr Ruche looked towards the window anxiously.

'Go and see what's happening, Pierre', Don Ottavio murmured.

Max and Sidney had been alone when the Short Stocky Guy arrived and started shouting at Sidney: 'So, Polly, you're not going to talk? You're doing this deliberately. See the state the boss is in because of you? If anything happens to him, I'll see to it you'll never talk again.' He reached out to grab Sidney.

'Leave him alone!' shouted Max.

'Shut it, kid!'

Sidney was flying round his head, screeching 'Fermat! Fermat!' Then she flew off. 'Come back', pleaded the Short Stocky Guy, terrified of what he had just done.

Max shouted after him, 'No! Sidney, come back, I promised...' but Sidney didn't hear. She flew high above the forest, cackling 'Fermat! Fermat!' and disappeared into the tropical sky, taking Grosrouvre's proofs with her.

'It's getting away!' shouted the Short Stocky Guy. 'Now it's talking, it'll tell everyone!' He took out his revolver and fired the shot that Mr Ruche and Don Ottavio heard. Max threw himself at the Short Stocky Guy. The man pushed him off, but it was too late. Sidney dropped like a stone into the trees around the house.

'You bastard! You killed him, you bastard!'

The Short Stocky Guy had seen the bird fall, and muttered under his breath 'You won't tell anyone anything now.' As he said it, he realized his mistake and his face went white. He started to tremble. If Don Ottavio found out, he would never forgive him.

He aimed the revolver at Max, who was still screaming, 'You killed him! You killed him!' The man panicked, his finger trembling on the trigger. He heard a noise behind him, but before he could turn, he had been knocked out cold.

Giulietta stood there with a baton in her hand.

'Are you all right, Max? He didn't hurt you, did he?'

'Thanks, Giulietta', said Max, getting to his feet.

He smiled. Giulietta thought he was smiling at her, but as he was getting up he had seen a bird just like Sidney rising from the trees where the parrot had fallen, and fly off into the forest.

Max didn't tell Mr Ruche what he had seen. It would be his secret. Mr Ruche decided that now that the bird was gone, there was no point in telling Max that Sidney was a female called Mamaguena, but he was surprised by the fact that the boy did not seem upset.

Mr Ruche went back to Don Ottavio's caravan to tell him what had happened. As soon as he got inside, he realized that Don Ottavio was dead. On the bedside table was a note. The door opened and Melissa rushed in, out of breath. She leaned close to Mr Ruche and said, 'A message for you from the hotel. They say you must phone Paris to Madame Perrette. She say it is very urgent.'

Urgent! Mr Ruche's heart skipped a beat. With Tavio dead and Sidney killed, what else could have happened? Giulietta offered to drive him to the hotel. He dialled the bookshop and Perrette answered. It was the middle of the night in Paris, and he had woken her.

'Perrette, it's Mr Ruche.'

'Has something happened to Max?' she asked.

'No, he's fine. Calm down. I was told that you had phoned me urgently, I thought something had happened to the twins.'

'No.'

'The bookshop?' he thought immediately of a fire.

'No, everything's fine. Just let me finish. I was reading the paper and I saw something.'

Mr Ruche listened as she explained, his face turning pale.

'Well, that's strange. That's really strange.'

Giulietta looked at him quizzically, and he held the receiver so she could hear as Perrette read from *Le Monde*. 'Fermat's Last Theorem proved! An English mathematician, Andrew Wiles, has proved one of the most famous theorems in mathematical history...'

Giulietta moved away from the receiver. As she did so, she said quietly to herself, 'It's just as well the boss died before he found out.' She smiled sadly. 'It would have killed him.'

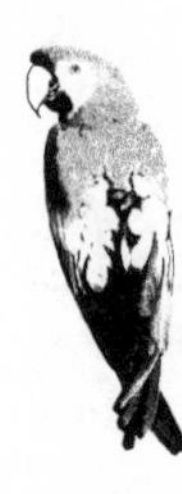

CHAPTER 26

Stepping Stones

A family reunion; a toast to Sidney & some explanations

The meal to welcome back Max and Mr Ruche was magnificent. When dessert was served, Perrette stood up and tapped her glass.

'I'd like to propose a toast. To us, now that we're all together again. All except Sidney, and we miss him.'

They raised their glasses. Even Max had been allowed a single flute of champagne.

'It's time to think about what we've learned', said Perrette as she sat down. 'Two of the three mysteries have been solved. Admittedly we didn't solve them, but they have been solved. The third – how Grosrouvre died – we don't know, though Mr Ruche has told us that Don Ottavio insisted he was not responsible for setting fire to the house. The only other possibilities are accident and suicide. From the information we have, there's no way for us to choose between them.

'There is, however, another problem: did Grosrouvre succeed in solving the two problems he set himself? I've been trying to find out something about this. There are two good reasons to believe that he didn't: the fact that Grosrouvre was an old man, and the fact that he was completely cut off from other mathematicians. I also found out about Andrew Wiles. While everyone says that a mathematician does his best work before the age of thirty-five, I found out that Wiles was over forty when he solved Fermat's Last Theorem. Grosrouvre would have been over sixty when he was working on it.

'The other problem was Grosrouvre's isolation. When mathematicians aren't in front of a blackboard or scribbling on a piece of paper, they spend a lot of their time at seminars and symposiums and international conferences. They meet weekly in maths departments or research centres. They talk, discuss their work, try out new ideas on their colleagues. How could a man working on his own in the middle of a forest, who has no direct contact with other mathematicians, succeed where the best minds in the world have failed? It's hard to believe, isn't it?'

Mr Ruche nodded, and Perrette went on: 'But I found out that, although he was employed in a university, Andrew Wiles didn't participate in a single seminar or conference for seven years before he published his solution. He didn't publish articles in magazines. Some of his colleagues thought he had stopped doing research work. So, he succeeded in solving Fermat's Last Theorem on his own. His only contact with other mathematicians was through reading mathematical journals.

'What about Grosrouvre? He had the Rainforest Library, which is certainly full of the greatest works in the history of maths, but it also contains a lot of books published recently. Now, books are usually published after discoveries have been made. Mathematicians usually first publish their work in journals – that's how they lay claim to their discoveries.'

'Because most scientists don't try to keep their discoveries secret, like Grosrouvre', said Lea.

'That's true, but remember that Wiles worked in total secrecy for seven years. He didn't publish any intermediary results. No one had read a single line of his work on the subject before he published.'

'At least he published it!'

'Let's get back to Grosrouvre. He subscribed to most mathematical journals. Grosrouvre may have been cut off from the world, but he knew what was happening in the world of mathematics. He may have been a couple of months behind, but he was not completely isolated, so there is no reason to think he couldn't have succeeded.'

Jon-and-Lea believed that Grosrouvre hadn't solved the theorems. Mr Ruche was in two minds. When they had started their research, he was convinced that Grosrouvre had solved the problems, but, as time went on, he began to doubt that his old friend was capable of such a feat. Max didn't care.

Perrette went on: 'At first, I thought it wasn't important whether Grosrouvre had solved the problems or not, because truth isn't important in a myth. But when you were in Syracuse, I changed my mind completely and took the mathematical point of view. In mathematics, truth is essential. I decided it was important to know whether Grosrouvre had solved the theorems.'

The doorbell rang.

'Who could that be at this time of night?' asked Mr Ruche.

Jonathan went downstairs to answer the door, and came back with Albert and Mr Habibi.

'We came to celebrate your homecoming.'

Lea offered them a drink. Max left the table, looking sadly at the perch where Sidney had sat. He went up to his room.

Perrette picked up where she had left off. She explained the two piles of journals in the last crate and the articles she had found underlined. 'I thought there might be a clue there somewhere, something that Grosrouvre was trying to tell us. But how could I find out? I thought about asking the guide from the Palace of Discoveries, you remember? I copied out the titles of the articles Grosrouvre had underlined into two lists, one for each pile of journals. I went and asked the guide if there was any link between the articles and Andrew Wiles' proof. He was surprised, but agreed to help.

'He phoned me the next day and I went to see him. He said, "Every single one of the articles in this list" – he showed me the longest list – "contains results or methods that Wiles used in his proof. Imagine the final proof as a river that everyone says cannot be bridged. The list of articles you've given me are the stepping stones – all of them! We know they are all there because Wiles used them and he made it to the other side." '

Perrette was excited now. 'Grosrouvre had all the stepping stones he needed, but did he make it to the other side? It's possible – probable, even. But he could have fallen into the river along the way. There's nothing to say he proved Fermat's Last Theorem, but...we met up again and I asked him about the other list.'

'What did he say?' Jonathan was excited now.

'All the articles on the second list were about Goldbach's conjecture', said Perrette. 'The underlined articles were the stepping stones for Grosrouvre to cross the stream of gold.'

It was a statement, not a question.

The lights went out.

Habibi and Albert burst out laughing, for the door opened and Max came in lit up like a Christmas tree. He walked slowly, carrying a cake covered with a veritable forest of candles, singing 'Happy Birthday!'

Max walked over to Mr Ruche carrying the cake. Mr Ruche smiled – he had outlived Diophantus, al-Khayyām and Grosrouvre. He had made it to his eighty-fifth birthday.

In his pocket, there was a note from Don Ottavio that he had brought back from Manaus. Mr Ruche decided not to mention the note. It would be his secret.

Epilogue

The conference of the birds

Night was falling. All over the world, animals were making their way to watering holes to drink, and in the forests noises were dying to a murmur. In the middle of a clearing in the Amazon a raucous voice screeched out.

Perched high in the branches of a cocoa tree, Mamaguena, alias Sidney, was talking. She didn't repeat, she didn't report, she recounted. More precisely, she proved...

All around the branches were filled with birds. Hundreds of them, of every colour, shape and size. They sat silently, attentively. On a branch opposite sat a beautiful blue ara, looking directly at Mamaguena.

In the respectful silence, Mamaguena explained to the conference of the birds each of the long proofs Grosrouvre had confided to her. The moon rose in the sky, and illuminated the clearing. Suddenly, one of the audience began to twitter, flapping its wings noisily. All heads turned to him disapprovingly, but he kept up the racket. Mamaguena looked worried. Could it be that the heckler had spotted something, some small thing in Grosrouvre's proof of Goldbach's conjecture, that was wrong...

Cast of characters